SpringerBriefs in Education

We are delighted to announce SpringerBriefs in Education, an innovative product type that combines elements of both journals and books. Briefs present concise summaries of cutting-edge research and practical applications in education. Featuring compact volumes of 50 to 125 pages, the SpringerBriefs in Education allow authors to present their ideas and readers to absorb them with a minimal time investment. Briefs are published as part of Springer's eBook Collection. In addition, Briefs are available for individual print and electronic purchase.

SpringerBriefs in Education cover a broad range of educational fields such as: Science Education, Higher Education, Educational Psychology, Assessment & Evaluation, Language Education, Mathematics Education, Educational Technology, Medical Education and Educational Policy.

SpringerBriefs typically offer an outlet for:

- An introduction to a (sub)field in education summarizing and giving an overview of theories, issues, core concepts and/or key literature in a particular field
- A timely report of state-of-the art analytical techniques and instruments in the field of educational research
- A presentation of core educational concepts
- An overview of a testing and evaluation method
- A snapshot of a hot or emerging topic or policy change
- An in-depth case study
- A literature review
- A report/review study of a survey
- An elaborated thesis

Both solicited and unsolicited manuscripts are considered for publication in the SpringerBriefs in Education series. Potential authors are warmly invited to complete and submit the Briefs Author Proposal form. All projects will be submitted to editorial review by editorial advisors.

SpringerBriefs are characterized by expedited production schedules with the aim for publication 8 to 12 weeks after acceptance and fast, global electronic dissemination through our online platform SpringerLink. The standard concise author contracts guarantee that:

- an individual ISBN is assigned to each manuscript
- each manuscript is copyrighted in the name of the author
- the author retains the right to post the pre-publication version on his/her website or that of his/her institution

Aaron M. Ellison · Manisha V. Patel

Success in Navigating Your Student Research Experience

Moving Forward in STEMM

Springer

Aaron M. Ellison
Sound Solutions for Sustainable Science LLC
Boston, MA, USA

Manisha V. Patel
Sound Solutions for Sustainable Science LLC
Boston, MA, USA

ISSN 2211-1921 ISSN 2211-193X (electronic)
SpringerBriefs in Education
ISBN 978-3-031-06640-5 ISBN 978-3-031-06641-2 (eBook)
https://doi.org/10.1007/978-3-031-06641-2

This Springer imprint is published by the registered company Springer Nature Switzerland AG
The registered company address is: Gewerbestrasse 11, 6330 Cham, Switzerland

Preface

Research is at the core of scientific inquiry. Engaging students in research as early as possible in the course of their education gives them opportunities to critically examine and, ideally, affirm their commitment to a career in Science, Technology, Engineering, Mathematics, or Medicine (STEMM); contributes to diversification of the STEMM workforce; and helps build networks of the next generation of scientists and STEMM educators. Indeed, many career scientists credit their participation in undergraduate research experiences with cementing their interest in STEMM fields and launching them onto successful career trajectories in academics, government agencies, nonprofit organizations, and the private sector.

Each student experiences research differently, but there are common approaches and practices that contribute to the success of every student researcher. In this book, *Success in Navigating Your Student Research Experience: Moving Forward in STEMM*, and its companion, *Success in Mentoring Your Student Researchers: Moving STEMM Forward*, we draw on our combined more than 50 years of experience and work as undergraduate researchers, research mentors, program managers, and undergraduate research program leaders to identify these common approaches. This book brings together in one place a set of best practices that students can use to find, engage with, and succeed in intensive, experiential research outside a college classroom.

The undergraduate research experiences we focus on in this book normally are full-time, usually paid, summer, semester, or year-long positions such as "REU" positions (in the USA) or "internships" (worldwide), but may also be a senior thesis project or a semester- or year-long independent study "class" for credit. As we discuss in Chap. 1, all of these differ substantially from the variety of well-established modes and mechanisms of active learning, such as inquiry- and project-based learning-by-doing, flipped classrooms, and place-based study.

Actually doing undergraduate research is a multifaceted process that involves preparation, undertaking the research itself, and further developing a STEMM career. These three activities are mirrored in the three parts of the book, each of which is divided into three chapters. Most chapters also have one or more Text Boxes and Vignettes. The Text Boxes concisely illustrate key ideas or tools, or suggest a set of questions or prompts to guide students through their research experience. The Vignettes are short, first-person narratives solicited through an open call on social media to current and former students about their own experiences preparing for, doing, or building on their undergraduate research.

In Part I, we start by discussing the many different reasons for doing research in STEMM, why it is important to start doing it as soon as possible as an undergraduate, and how to go about finding undergraduate research opportunities (Chap. 1). For many students, especially those from minoritized groups that are still underrepresented in STEMM fields or who are the first in their family to attend college, finding an undergraduate research experience and navigating the oft-times complex application process can be off-putting and intimidating. Thus, in Chap. 1, we provide tools for students to use to work through their needs and expectations for doing research. We then provide extensive guidance on the range of available resources that they can use to find a research experience that matches their needs and expectations.

In Chap. 2, we work through the application process in detail, highlighting strategies for success and pitfalls to avoid along the way to obtaining a position. And once a position is secured—a process that routinely requires applying to multiple programs and either choosing among multiple offers or coping with rejection—there is a relatively short time to prepare for it. In Chap. 3, we emphasize that key elements of preparation not only involve learning about the research itself but also sweating the logistical details needed to ensure, for example, that any travel needed to get to the research experience is booked, housing is secured, and that paychecks arrive on time.

In Part II, we dive into the details of how doing undergraduate research differs substantially from what students may have experienced in their college classes, even those where active learning is a central component. We consider the undergraduate research experience itself as a "three-legged stool," whose legs are research, education, and community. The balance among these legs helps to ensure success in research and careers in STEMM.

First among equals, an undergraduate research experience gets students doing real research (Chap. 4). Second, an undergraduate research experience provides a range of opportunities to learn about the broader context of a research topic and learn the professional skills that all scientists need to be successful in STEMM. These skills include writing research proposals, navigating permitting requirements, communicating results to a range of audiences, identifying career trajectories, and building professional networks (Chap. 5). Finally, undergraduate researchers become part of a much larger and supportive community of STEMM professionals. In Chap. 6, we discuss strategies student researchers can use to successfully build and become part of their own STEMM communities.

Finally, the undergraduate research experience does not end when the summer, term, year, or program itself ends. The three chapters in Part III focus on how to take advantage of the new short- and long-term opportunities that doing research opens up. We provide guidance on how undergraduates can share their research with a variety of audiences through informal channels, in more formal publications, and at professional meetings and conferences (Chap. 7). Our discussion of subsequent opportunities that students have to expand their research further naturally leads to a consideration of further employment, graduate school, and STEMM careers (Chap. 8). In the best of all cases, the students themselves become mentors to the next generation of students. We conclude the book by encouraging students to "pay it forward" by becoming mentors to the next cohort of undergraduate researchers (Chap. 9).

Like doing research itself, writing and publishing a book is a team effort. Our ideas about mentored undergraduate research have been refined by our work with hundreds of students and dozens of our colleagues—including administrative and support staff, graduate students and post-docs, senior researchers, and faculty members—at the institutions where we have worked (Swarthmore College, Mount Holyoke College, Rutgers University, the University of Vermont, and Harvard University) and with whom we collaborate through various professional networks (BIO-REU, ESA-SEEDS, LTER, OBFS, and UFERN). We are deeply grateful to Beth Fischer (Assistant Professor in the School of Education at the University of Pittsburgh) and Michael Zigmond (Professor Emeritus of Neurology, Psychiatry, and Neurobiology at the University of Pittsburgh) for teaching us the importance of "survival skills" in STEMM through their workshops and inviting Aaron Ellison to participate in their "trainer-of-trainers" conferences in the 1990s; and to Brad Rose (Brad Rose Consulting) and Andrew McDevitt (Lecturer in Biology at the University of Colorado, Denver) for teaching us how to evaluate undergraduate research programs, and then working with us and undergraduate researcher Relena Ribbons (now Assistant Professor of Geosciences at Lawrence University) to evaluate the Harvard Forest Summer Research Program in Ecology. Last, we thank Claudia Acuna, our editor at Springer Nature for taking on this book and seeing it through to publication; two anonymous reviewers who provided useful comments on the book proposal and the final manuscript; and Rajan Muthu and his production team in Chennai who wrangled our manuscript into its final form.

Over the last three decades, our work as undergraduate research mentors and undergraduate research program co-directors has been supported not only by our home institutions but also by a range of competitive awards. For support of individual undergraduate researchers, these have included grants from the DOE (award no. DE-FG02-08ER64510); the Ellen P. Reese Fund at Mount Holyoke College; the HHMI; the Massachusetts Natural Heritage and Endangered Species Program; The Mellon Foundation; the US National Institute of Climate Change Research (NICCR); NSF (award nos. BSR 9107915; DEB 9253743, 9741904, 9805722, 9942207, 0115145, 031361, 0235128, 0422750, 0520792, 0528625, 0541680, 0722588, 0816508, 0902592, 1025362, 1110434, 1144056, 1136646, 1518653); and the Orchards Golf Course in South Hadley, Massachusetts. For support of undergraduate research programs, these have included grants from the HHMI (award no. 71196-505002); the NSF (award nos. DBI 0330605, 0422745, 0452254, 0520794, 0618448, 0812997, 0930516, 1003938, 1111158, 1224437, 1239937, 1341122, 1446653, 1459519, 1535283; ACI 1450277); NASA (NNX10AT52A); and the Sherman Fairchild Foundation.

Finally, we especially acknowledge the influence and impact of two permanent program directors at NSF—Sally O'Connor and Sonia Ortega—whose complementary visions for undergraduate research have shaped our own and who have been instrumental in advancing undergraduate research programs across the USA and around the world.

Boston and Greenfield, USA
Singapore
January 2021

Aaron M. Ellison
Manisha V. Patel

Contents

Acronyms

AANAPISI	Asian-American- or Native American and Pacific Islander-Serving Institutions (see MSI).
ANNHI	Alaskan Native- or Native Hawaiian-Serving Institutions (see MSI).
BIPOC	Black, Indigenous, or People of Color.
BMES	Biomedical Engineering Society.
CIMER	The Center for the Improvement of Mentored Experiences in Research, an organization that has developed a number of platforms for assessing outcomes of undergraduate research.
CRediT	Contributor Roles Taxonomy, a system that provides clear guidelines for maintaining integrity and transparency in determining and asserting (co-)authorship of research papers.
CUR	Council on Undergraduate Research.
CV	Curriculum Vitae: a complete history of one's education, experiences, and accomplishments.
DAAD	Deutscher Akademischer Austauschdienst (German Academic Exchange Service).
DOD	US Department of Defense.
DOE	US Department of Energy.
EPA	US Environmental Protection Agency.
FAIR	Findable, Accessible, and Interoperable and Reusable. FAIR refers to a set of guidelines that encapsulate current best practices for responsible and ethical data management.
HBCU	Historically Black Colleges and Universities (see MSI).
HHMI	The USA-based Howard Hughes Medical Institute.
HSI	Hispanic-Serving Institutions (see MSI).
IPSE	International Society for Pharmaceutical Engineering.
LTER	The Long-term Ecological Research program supported by the NSF.

MSI	Minority-serving Institutions: USA-based colleges and universities that have an historical origin in serving minoritized individuals or now have a large percentage of students from minoritized groups. Types of MSI include AANAPISI, ANNHI, HBCU, HSI, NANTI, PBI, and TCU.
NANTI	Native American Non-Tribal Institutions (see MSI).
NASA	US National Aeronautics and Space Administration.
NGO	Non-governmental Organization.
NIH	US National Institutes of Health.
NSF	US National Science Foundation.
NSPE	USA-based National Society for Professional Engineers.
OBFS	Organization of Biological Field Stations
PBI	Predominantly Black Institutions (see MSI).
RCR	Responsible Conduct of Research: the umbrella term for promoting the goals and objectives of scientific inquiry, and for creating and nurturing a collegial and inclusive environment that allows diverse scientists to work together toward agreed-upon, common goals.
REU	Research Experience for Undergraduates, a cross-directorate program of the NSF.
SACNAS	USA-based Society for Advancement of Chicanos/Hispanics and Native Americans in Science.
SALG	The Student Assessment of Learning Gains, a survey instrument that is commonly used to evaluate the outcomes of undergraduate research experiences.
SEEDS	Strategies for Ecological Education, Diversity, and Sustainability. SEEDS is a program of the Ecological Society of America that aims to diversify and advance the profession of ecology through opportunities that stimulate and nurture the interest of underrepresented students to participate and to lead in ecology.
STEM	Science, Technology, Engineering, or Mathematics.
STEMM	Science, Technology, Engineering, Mathematics, or Medicine.
TCU	Tribal Colleges and Universities (see MSI).
UFERN	Undergraduate Field Experiences Research Network.
URE	Undergraduate Research Experience.
UROP	Undergraduate Research Opportunity.
URSSA	The Undergraduate Research Student Self-Assessment survey.
USDA	US Department of Agriculture.

List of Boxes and Vignettes

Part I
Preparing for Your Research Experience

In the Part I of this book, we discuss the many different reasons for doing research in science, technology, engineering, mathematics, or medicine (collectively and colloquially known as STEMM); why it is important to start doing research as soon as possible during your undergraduate years; how to go about finding and applying for opportunities to learn research by doing it; and how to prepare yourself to hit the ground running from day one.

Chapter 1
Finding a Research Experience

Abstract This chapter discusses the importance of an undergraduate research experience in science, technology, engineering, mathematics, or medicine (STEMM) and how it differs from lecture-based or "active" classroom learning, and "independent study" for credit. Different types of research experiences and internships can range from the traditional one-student-one-mentor experiences to participation in larger, more structured undergraduate research programs that may have teams of multiple students and mentors. Guidance is provided on the range of available resources that students can use to find a research experience that matches their needs and expectations.

Doing original research is a defining moment for anyone aspiring to further study or to have a career in science, technology, engineering, mathematics, or medicine (STEMM), and for everyone interested in learning more about how science "works." But what exactly is "original research?" How do you know if and when doing research is right for you? And if you decide that it is the right time to do research, how do you find the best opportunity for you to do it? Read on!

In this first chapter, we discuss the importance of a research experience in STEMM and how it differs from lecture-based or "active" classroom learning, and "independent study" for credit. We delve into the different types of research experiences and internships. These can range from the traditional one-student-one-mentor experiences to larger, more structured undergraduate research programs, some of which have teams of multiple students and mentors. We also provide guidance on the range of available resources that students can use to find a research experience that matches their needs and expectations.

There are two Text Boxes and a Vignette in this chapter. Box 1.1 encourages you to reflect about whether a research experience is right for you. Box 1.2 provides links to a wide range of Internet resources for finding undergraduate research experiences. Vignette 1.1 describes how the authors got their start in research.

A. M. Ellison and M. V. Patel, *Success in Navigating Your Student Research Experience*, SpringerBriefs in Education, https://doi.org/10.1007/978-3-031-06641-2_1

1.1 What Is Research?

> Research is any creative systematic activity undertaken in order to increase the stock of knowledge, including knowledge of people, culture, and society, and the use of this knowledge to devise new applications.
>
> —(OECD 2007)

If you're reading this book, chances are you had your first experience with what was presented to you as scientific research in a lab section of a general science course such as an introduction to astronomy, biology, chemistry, geology, or physics. During lectures, you were presented with a set of more-or-less connected facts and figures you needed to memorize for an exam.

When you got to your lab section, you were assigned a question, perhaps read a chapter or two in a textbook or a journal article, learned and worked through a series of hands-on methods, collected and analyzed the data, and wrote up a summary of the whole process as a lab report. Such exercises are designed to introduce some background information and to teach a set of methods and the process of collecting and analyzing data—the "scientific method."[1] Such problem-solving methods introduced you to basic questions and specific lab methods and techniques, but these are only the first of many stepping-stones into what practicing scientists think of as research (Fig. 1.1).

In later courses, or perhaps later in the same course, you might have been asked to think more deeply about a topic covered in the class, to discuss it with other students in the same class, or to come up with your own question to ask. This "active" learning (e.g., Barkley et al. 2014) or "inquiry-based" learning (e.g., National Research Council 2000; Rissing and Cogan 2009), both of which are now common in "flipped" classrooms (e.g., Herreid and Schiller 2013), is much more engaging. Most importantly, when you participate in inquiry-based learning, you retain much more of the information and develop a deeper understanding of the concepts and methods

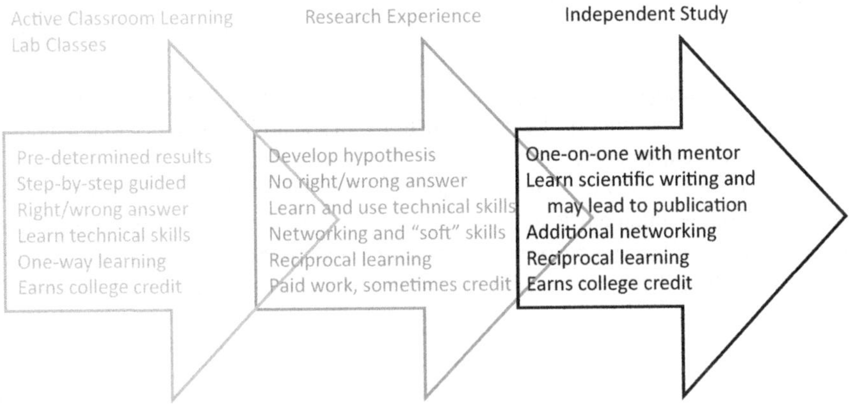

Fig. 1.1 Three stages of undergraduate learning and research

that you had to learn to answer your own questions. But inquiry-based learning rarely includes research.

Research is much more than lab exercises or active- or inquiry-based learning. Research starts with asking a question that you are interested in *but don't know the answer to*. In fact, it is unlikely that anyone knows the answer to the precise research question that you are asking! Research then proceeds through the process of figuring out how to answer the question, trying out a series of methods (some of which might even work), collecting and analyzing the data, and then deciding if the data really can be used to suggest *an* answer to the question you started with. The process itself is more important than the answer. Research is about pushing our collective understanding to the next level, and because the "answer" to your question wasn't known when you asked it, you're not going to find the answer in books, articles, or on the Internet. It may even be hard to know if the answer you reached is "right" or not. And finally, while you were working on answering your initial question, many more questions came up, some of which might already have answers while others will lead you in new and different directions. Research is an endless process that can be simultaneously exhilarating and intimidating, exciting and boring, or rewarding and frustrating.

1.1.1 What Is Undergraduate Research?

Undergraduate research is an *original investigation which results in an intellectual or creative contribution to a field of study* (Wenzel 1997). It is the next step in learning and understanding how to do scientific research (Fig. 1.1). Taking this step by participating in a research experience can put you on a track towards undergraduate thesis research, graduate or professional school, and a job or career in STEMM. By doing undergraduate research, you can start right now learning the process of doing scientific research and figuring out if this might be something you want to pursue as a career. It can also refocus your interests and drive you to make a personal or professional change in your longer-term goals. This chapter and the next two chapters provide a guide for getting you out of the classroom and into your first research experience.

We note that in different settings, institutions, and countries, an undergraduate research experience ("URE") may have different names or titles. These include internship, research assistantship, research experience for undergraduates ("REU"), and undergraduate research opportunity ("UROP"), among others. As long as the actual activities include the opportunity for you as an undergraduate student to do original work that results in an intellectual or creative contribution to a field of study, we include it in what we refer to throughout this book as an "undergraduate research experience."[2]

1.2 Understanding the Different Types of Undergraduate Research Experiences

Research is about asking questions and framing hypotheses, collecting data to answer the questions and test the hypotheses, interpreting the data, and using those interpretations to ask new questions, re-frame your hypotheses, and further and deepen our understanding of the world around us. Unlike in a lab exercise, there is no right or wrong answer in research. There are just data, interpretations, more questions to be asked and explored, and new hypotheses to be tested. As with research itself, there are no right or wrong outcomes from undergraduate research experiences. Just as outcomes from research experiences can be different, their structures and implementations differ, too. Before turning to their differences, we highlight commonalities among different types of undergraduate research experiences.

1.2.1 Common Components of Undergraduate Research Experiences

There are three key features of an undergraduate research experience that distinguish it from lab exercises and active classroom learning methods that simulate the research process (Fig. 1.1).

First, in an undergraduate research experience you will learn to develop hypotheses and research questions that build on previous work *but for which no one knows the outcome or answer*. Thus, you, your fellow students, and your mentor(s) will all be thinking, learning, and struggling together. You will also be taught (or perhaps just shown) techniques and be expected to understand and use them right away. We discuss the practices of undergraduate research in detail in Chap. 4.

Second, in addition to learning new technical skills, you will also learn to use them and, when needed, modify them for the tasks at hand. As a member of a research team, you will also have opportunities to network with other students and professional scientists, and learn many of the other important skills that scientists use every day but rarely think to teach others. These skills include: ensuring your research practice meets the highest ethical standards; effectively reading the scientific literature; writing abstracts and grant proposals; honing your résumé for specific positions and careers; and many others (for more in-depth discussion of the educational and skill-building aspects of undergraduate research experiences, see Chap. 5).

Third, most undergraduate research experiences are funded: you will receive an hourly wage or stipend, and in many cases your housing, food, and travel will also be covered. Some universities will also provide college credit for undergraduate research experiences, but this is less common. If you want to get college credit for a research experience, find out if credit is possible before you apply for the experience. You will want to ask both your home institution and, if it's different, the institution where you're applying for the research experience.

1.2.2 *Different Types of Undergraduate Research Experiences*

There are two broad types of undergraduate research experiences: individualized opportunities, such as working in a researcher's lab or research group, and formal programs that bring together a cohort of students in one or more research groups. Both provide research experiences, but there are salient differences between them.

Individualized research experiences

A solo or one-on-one experience generally takes leg-work on your part to find (Vignette 1.1). You will most likely find individualized research experiences at your home institution, but few professors actually advertise undergraduate research positions. They might announce these positions in a class, but they will more commonly expect that interested students will approach them about working in their research group. If you need to get paid, you will have to ask for a paid job or internship. Although you may already know and be comfortable in the general location and broader cultural environment where you're going to school, you may not be familiar with research culture or uncomfortable with the thought of "cold-calling" a professor and asking for an opportunity. Don't be shy, though. Even though you may not get a position, you may get a pointer to other opportunities (Vignette 1.1).

Vignette 1.1 How the authors got their first research experiences

Contributed by Manisha V. Patel and Aaron M. Ellison

Manisha: I was a junior in college when I realized that I needed a job in my major to get experience. From a search of the job board, I applied for and got a part-time position as seasonal wildlife technician in the Fish & Wildlife department. While working there, I heard about an opportunity with one of the researchers for an assistant in a wildlife study. I was really interested, but lacked the skills and confidence to speak with him in person. I wrote a cover letter and résumé and left it on his desk. Although I did not get that position, he passed my résumé on to a colleague looking for summer help on another research project. I was offered that job, but accepted it with some hesitation. For this job—my first experience in research—I lived in a cabin in a state park because the work was too far from home for a daily commute but close enough that I could drive home every weekend. Being a novice, I learned some methods of scientific data collection but lacked the skills to ask and learn more about the research. It was the first stepping-stone for me, and set me on a path to a number of different field and lab-tech jobs (in the same professor's lab but working on different research projects with different members of the lab) that, in hindsight, combined into a more complete research experience. As I grew more confident, I asked more questions and became proactive in learning how to do scientific research.

Aaron: I was a sophomore in college and needed a job to pay for anything beyond tuition, room, and board. I was very interested in science, but lacked any specific direction. I found, applied for, and was hired as a runner/tech in the surgical pathology lab of the university's hospital. I discovered that I really hated pathology, and after two weeks, returned to the campus job board. I found a job sorting specimens stored in alcohol in the basement of the university's museum. It was quiet, pleasant work, and in the spring, I asked the professor who had hired me if I could keep doing it over the summer. He had no money for that, but with a phone call to his budget office (while I was sitting in his office), he was able to reallocate funds meant for a new lab refrigerator to hire me for the summer as a field assistant. I quickly signed on to what turned out to be two months of driving a battered pickup with a microscope and fiber-optic light hooked up to a portable generator into farms and forests across the region in search of slime molds in cow patties and piles of leaves and soil. I never found any slime molds that way, but the experience hooked me into ecology. That summer, I did come up with an alternative field-sampling method that led us to build up a good collection of slime mold species in the lab, and I continued working with the same professor in his lab throughout my junior and senior years.

Once you start working with a professor and their research group, you will be plunged almost instantly into the hypothesis-driven research being done by the professor and any post-docs, grad students, or undergraduates already working in the group. These folks are part of your new professional community (see Chap. 6 for more discussion about building your own community in STEMM). Since there are no right or wrong answers, everyone is grappling for direction and learning along the way. Even though everyone in the research group is working towards a common research goal, they also each have a different set of priorities and expectations. You may find these differences make for a greater diversity of perspectives, or you may feel lonely because no one else appears to be "in the same boat" with you.

It may also seem like everyone else in the research group is already proficient in reading scientific literature, analyzing data, writing and presenting proposals results, but how do you learn these skills since they're not part of your job and there aren't obvious courses to take? In an individualized research experience, you'll need to either pick up these (and other) skills on your own (and on your own time) or ask for help (but be prepared for no one to have the time, knowledge, or skills to teach you). You are more likely to get formal education in a range of non-technical skills in larger or more structured undergraduate research programs (Chap. 5).

In an individualized research experience, you might have the opportunity to develop your own research project. If you're the only undergraduate—or even the only other person—working in the research group, your mentor (the professor, post-doc, or grad student) may be able to give you the individualized attention and support your need to develop an independent project. But others in the research group may just be looking to you for an extra set of hands. If you want this to be a full research experience, you should be pro-active and upfront in conversations about your expectations (for both yourself and your potential research mentor). We discuss the importance of clear communication in more detail in Chaps. 3 and 4.

Undergraduate research programs

Undergraduate research programs incorporate multiple aspects of science. Although doing research is still their main focus, undergraduate research programs also provide more formal education and opportunities to learn professional skills and bring students into broader research communities.

Undergraduate research programs advertise opportunities broadly (see Sect. 1.4, below) and have a formal application process. Programs usually recruit regionally or nationally, so these could be good opportunities for you to discover another college, region, or culture. Virtually all undergraduate research programs pay a good stipend (above minimum wage) and many include housing, meal plans, and travel to and from the program. Some include short-term health insurance or worker's compensation coverage.

But going somewhere else for a research experience can also be a deterrent. Other responsibilities and obligations may mean you can't go far from home or school. Don't despair, though. It's quite possible that your home institution has one or more undergraduate research programs, or a nearby government lab or agency, non-profit, or scientific or research-based company (e.g., pharmaceutical development or high-tech company) has programmatic opportunities for undergraduate research. Online ("virtual") undergraduate research programs are also increasingly common.

When it comes to the research itself, undergraduate research programs can vary a lot. In some, students are embedded into individual research labs with predefined goals, while in others you may be working with cohorts of similarly enthusiastic students with multiple research mentors. You may have a lot of flexibility in defining both the big picture and the fine details for your research question or these may be set in large part by the individual mentors or the overall program. The good news is that this information should be available on the program's website. This can be especially helpful if you're not comfortable approaching researchers directly for an individualized research experience.

Because they host and support many students (usually from many different places and diverse backgrounds), undergraduate research programs also are a great place to meet and work with a variety of students with similar interests yet different perspectives. For many student researchers, such programs are their first opportunity to become part of a network of a diverse community of researchers (Chap. 6). The multiple mentors associated with undergraduate research programs can also provide opportunities to meet different scientists and be exposed to different academic disciplines.

Undergraduate research programs also provide multiple avenues to learn the non-technical skills that are so crucial to ultimate success in science. You can expect to participate in some combination of workshops, study groups, information sessions (for example, on applying to graduate or professional schools, non-academic career options), and networking opportunities. Each component is taught in a formal way and then you will have multiple opportunities to practice these skills by incorporating them into different aspects of your research experience (Chap. 5).

Because these educational components are integrated into undergraduate research programs, you may have less time to be doing the actual research itself. The main goal of these programs is to provide students with a well-rounded research experience, with equal emphasis on the research and the experience. If you find you aren't enjoying the particular research topic you started with, you may have more opportunities in a structured program than you would in an individualized research experience to learn (and maybe find that you like) other topics. There might also be other aspects to the experience that you find interesting and enjoy, including understanding the bigger picture of scientific research and becoming part of another community and support system.

1.3 Knowing Yourself: Do You Need a Research Experience?

As an undergraduate, your future probably looks like getting passing grades this term, choosing your courses for next term, or picking a major. You might be interested in a specific area of STEMM, but you might also be worried that STEMM isn't interested in you. For example, you may have done well in science classes in high school, but you're having less success in your college science courses and more success in other fields of study. Maybe you took an introductory course but you're unsure about how to advance in the field and the department doesn't seem inclusive and open to new people. Or you're doing really well in your science courses, but you feel like an imposter among your classmates who you assume are smarter than you are. All of these feelings are real and shared but unvoiced by most of your classmates. While you shouldn't let anyone but yourself say "you can't [or shouldn't] 'do' science or be a scientist" you can be sure that the road ahead will not be the same as the paths walked by other students who have come before you or those who will come after you. Pursuing a career in STEMM is an opportunity to create and be on your own road.

Box 1.1 poses a series of reflective questions that can help you think about whether you want a research experience now and clarify your short-term goals in scientific research.

Box 1.1 Self-reflection: Is a research experience right for me?

Now is a good time to start thinking about whether you want or need a research experience. Answering the following questions can help you decide.

- Is now the time? The answer is not a simple yes or no. Rather, ask yourself "If I want to do this, what are factors I need to consider?"
- Am I ready to push myself outside of my comfort zone?
- Would a research experience add to my résumé and help me take the next step on my career path?
- How do I learn best? Through observing (including reading) or practice?

If the answers to these questions lean you towards pursuing a research experience, then this next set of questions will help you narrow down which type of experience you might look for.

- Do I need to have a paying job on a regular basis?
- Am I okay going far away or do I want or need to stay close to home or school?
- What type of support do I want from the research community? For example, do I want to work one-on-one with a more senior researcher (mentor) or do I want to work in teams with other students?
- Do I learn better working informally at my own pace or in a structured internship or program?
- Would I thrive in a more or less diverse environment?

A few aspects of undergraduate research are of particular importance. First, the environment—both the physical location and the community of people—matters a lot. If you decide to pursue a research experience, you want to be able to fully commit to it. Really consider your priorities, obligations, and work-life balance (see also Sect. 3.1 in Chap. 3). Once you know your priorities, you can begin looking for opportunities that fit your needs and desires. Keep in mind that each research experience has its own context, structure, and priorities, but that undergraduate-research mentors are approachable and can usually be flexible in meeting your needs.

If a research experience seems too daunting or you can't commit fully to it right now, you can still learn more about how research works by finding part-time work, as, for example, a lab or field assistant or technician in an active research group. Such an initial, partial immersion in research could give you the additional confidence you need to pursue an undergraduate research experience in the near future.

1.4 How to Find Research Experiences

If you've gotten to this point in the chapter, you've decided that you're ready to find a research experience and start doing research. You've also probably thought about whether you want to work one-on-one with a more senior researcher in an individualized research experience or look for a more structured undergraduate research program (Sect. 1.2, above). Either way, you've got work to do.

After deciding that this is the right time for a research experience (Box 1.1), you should consider which broad STEMM fields are most interesting to you. Most colleges and universities have undergraduate biology, chemistry, computer science, mathematics, and physics departments. Astronomy, environmental science, engineering, and statistics may have their own departments or be programs within or between (i.e., "interdisciplinary") existing ones. Medical research can be found in medical schools, some research institutes, national laboratories (such as the US National Institutes of Health or those supported worldwide by the UK's Wellcome Trust), and the private sector (e.g., as divisions within pharmaceutical, medical technology, or drug-delivery companies). In any of these fields and their subfields, there are opportunities for lab- and field-based research or computational and theoretical work, and for research that involves collaborations across disciplines.

As we discussed earlier in the chapter, individualized research experiences tend to be poorly (if at all) advertised, mentioned in classes, or posted by internship offices. They can sometimes come about through conversations or via informal networks, or created after discussions with, or referrals from, senior researchers, professors, or other professionals who can create opportunities from available funds. Although both of us started our pathways into science doing individualized research experiences (Vignette 1.1), the proliferation of undergraduate research programs in the last 20 years has made it easier to explore and participate in a much greater diversity of research experiences while learning valuable skills and building communities and networks of like-minded researchers. If you are interested in going somewhere else to do research, are more comfortable working in groups or with formal structures, or are uncomfortable with the prospect of approaching professors or other more senior researchers, then undergraduate research programs are a better way to go.

1.4.1 Finding Undergraduate Research Programs

Undergraduate research programs have been developed most extensively in the United States. This may reflect the greater flexibility in undergraduate curricula in the USA and the deliberate creation of dedicated funding mechanisms that support undergraduate research opportunities. With some exceptions (notably the NASA Intern and Fellow Opportunities program), opportunities for undergraduate research in other countries tends to follow the individualized research model and can require more extensive conversations with senior researchers.[3]

The US National Science Foundation (NSF) supports research experiences for undergraduates through "REU Site" programs that bring together six or more students to do mentored research focused on a common scientific theme (e.g., applications of Big Data, protein synthesis, forest ecology). Any area of research supported by the NSF probably has at least one, and sometimes dozens of REU Sites (Box 1.2) and most REU Sites are based at research universities, basic research institutes, field stations, and marine labs.

REU Sites recruit students from across the USA[4] and actively seek out students who have not had opportunities to do research at their home institutions. Priority is often given to minoritized students or those attending minority-serving institutions (MSIs),[5] and students who are disabled, first in their family to attend college, are attending community (2-year) colleges, or have faced other barriers to engaging in scientific research.

There are also REU Sites at MSIs and liberal-arts colleges that draw their participants primarily from their home institutions or their surrounding regions. Undergraduate research in liberal-arts colleges and MSIs has a long history of success that is rarely acknowledged by larger "research" universities. The more than 500 institutions that are members of the Council on Undergraduate Research (CUR) reflect their deep and continuing commitment to "learning through research."

Many federal departments and agencies have their own research labs. Examples in the USA include the National Institutes of Health (NIH), Department of Energy (DOE), National Aeronautics and Space Administration (NASA), the Department of Defense (DoD), and the Department of Agriculture (USDA). The wide range of undergraduate research experiences offered by these and other federal agencies (Box 1.1) may lead you to subsequent internships and careers with them.

Large foundations (e.g., the USA-based Howard Hughes Medical Institute and the UK-based Wellcome Trust) and companies, including many in the technology and pharmaceutical sector, also provide research opportunities. These may be implemented as partnerships with colleges or universities that are similar to federally-supported undergraduate research experiences, or they may be advertised as paid internships. Some of these opportunities have full websites (Box 1.2), whereas others are advertised through professional (social) networks such as LinkedIn, Indeed or ResearchGate.

Finally, most scientific and professional societies have websites, list-serves, and online alert systems where field- and discipline-specific research opportunities are posted. In recent years, there has been increasing emphasis on, and coalescence of organizations dedicated to, greater diversity and inclusion in STEMM careers. These organizations can provide additional pointers and guidance for getting started in and sustaining STEMM careers (Box 1.2).

Box 1.2 Where to start looking for undergraduate research programs

- Biomedical Engineering Society (BMES)
- Council on Undergraduate Research
- DAAD German Academic Exchange Service
- DoD STEM Internships
- DOE Scholars Program
- Ecological Society of America's Strategies for Ecology Education, Diversity and Sustainability (ESA-SEEDS)
- International Society for Pharmaceutical Engineering (IPSE)
- NASA for International Students
- NASA STEM Engagement
- National Society of Professional Engineers (NSPE)
- NIH Summer Internship Program
- NSF REU Sites
- OneUSDA Internship Program
- Pathways to Science
- Society for Advancement of Chicanos / Hispanics and Native Americans in Science (SACNAS)

1.5 Take-Home Messages

✔ Start now learning the process of doing research through an undergraduate research experience.

✔ Know yourself and your priorities to understand what kind of undergraduate research experience is right for you.

✔ You have work to do to find your undergraduate research experience.

References

Barkley, E. F., Major, C. H., & Cross, K. P. (2014). *Collaborative learning techniques: A handbook for college faculty* (2nd ed.). San Francisco: Wiley.

Herreid, C. F., & Schiller, N. A. (2013). Case studies and the flipped classroom. *Journal of College Science Teaching, 42,* 62–66.

National Research Council. (2000). *Inquiry and the national science education standards: A guide for teaching and learning*. Washington, DC: The National Academies Press.

OECD. (2007). *Glossary of statistical terms*. Paris: Organization for Economic Co-operation and Development.

Rissing, S. W., & Cogan, J. G. (2009). Can an inquiry approach improve college student learning in a teaching laboratory? *CBE Life Sciences Education, 8,* 55–61.

Wenzel, T. J. (1997). What is undergraduate research? *Council on Undergraduate Research Quarterly, 17,* 163.

Chapter 2
Applying for a Research Experience

Abstract This chapter discusses the process for crafting a more formal (and usually online) application and interviewing for undergraduate research experiences. These are important skills to master because success rates for applications to undergraduate research experiences range widely. Some programs may accept more than half of those who apply, whereas admission rates to the most competitive undergraduate research experiences and programs can be well under 10%. Tips and advice are presented for obtaining strong letters of recommendation, interviewing for research experiences, and handling both acceptance and rejection at all stages of the application process.

Undergraduate research experiences range from individual internships set up through informal conversations and on-campus networks to programs that recruit students from national and international applicant pools (Chap. 1). "Applying" for the former may only require asking a more senior researcher (graduate student, postdoc, faculty member, or private-sector researcher), but even being able to ask starts with being in the "right" place at the "right" time, knowing the "right" people, and having the self-confidence to approach them and ask. Although these may seem like small and easy steps to an experienced researcher, they may appear insurmountable for students who can benefit the most from undergraduate research experiences: students of color in majority-white institutions, first-generation college students, and students with no previous research experience. For example, Johnson (2012) and Blackburn (2017) discuss how a "chilly" departmental or institutional climate (sensu Seaton 2011) can affect a student's willingness to even approach more senior researchers about STEMM research.

Ironically, it may be more comfortable for you or any student—and especially those students who need a research experience the most—to apply online for an undergraduate research experience at your home institution, a college or university, with the government, or in the private sector. You likely believe, with some justification, that applications in a pool are treated and reviewed fairly and equally. The truth of this belief varies among programs, but in STEMM communities, we are constantly working towards this ideal as mentors learn new skills to reduce conscious and unconscious or implicit bias[1] and work together with program directors

A. M. Ellison and M. V. Patel, *Success in Navigating Your Student Research Experience*, SpringerBriefs in Education, https://doi.org/10.1007/978-3-031-06641-2_2

to provide research experiences to those students who can benefit the most from them (McDevitt et al. 2016).[2]

In this chapter we discuss the process for crafting a more formal (and usually online) application and interviewing for undergraduate research experiences. We discussed more informal approaches in Chap. 1, and emphasize in Chaps. 6 and 7 the importance of building and using networks to find additional opportunities. There are three Text Boxes in this chapter. Boxs 2.1 and 2.2 illustrate what makes for a compelling cover letter to an application. Box 2.3 highlights pitfalls to avoid in navigating online application systems. We also share tips and advice for interviewing, and talk through ways to handle both acceptance and rejection from individuals whom you may ask for letters of recommendation, potential advisors and mentors, and undergraduate research programs. These are all important skills to practice and master. Success rates for applications to undergraduate research experiences range widely; some programs may accept more than half of those who apply, whereas admission rates to the most competitive undergraduate research experiences and programs can be well under 10%.

2.1 The Application

A typical application for an undergraduate research experience is a lot like an application for any other internship or job. The key parts include: a letter of interest or cover letter, résumé or "CV" (*curriculum vitae*),[3] transcripts (which may be optional), and strong letters of recommendation. Some applications may also require a personal statements or short essays. Because different programs and application systems will always have specific requirements, you should always be prepared to deal with their differences.

We present the components of an application in the next few subsections, ordered in the way that we and many of our students have found easiest to complete. However, everyone approaches an application in their own way, and crafting one is an iterative process. For example, you may choose to tackle what you think are the easiest parts first, but as you work on other parts of your application, you are likely to find that you want to re-visit and modify previously "complete" sections.

2.1.1 The Résumé

It's often easiest to start by putting together your résumé. It consists of a set of basic components (your name, contact information, education, work history); other pieces specific to particular disciplines or jobs (such as a professional summary, career objective, skills, certifications, publications, community or volunteer work); and names and contact information for one or more references. A résumé can be formatted in many different ways. Word processors such as Microsoft Word or The Document

Foundation's LibreOffice[4] have built-in résumé templates, and other examples can be found using any Internet search-engine. College career offices and undergraduate advisors also can give you useful advice on crafting your résumé.

No matter the format or layout, a good way to get organized is to put together your more comprehensive CV with its complete history of your education, experiences, and accomplishments. You can then use your CV as an "à la carte" menu from which to extract specific components and build the unique résumé that you need for each application. Although this will take a little more time, it will be well worth the effort. A unique, carefully constructed résumé demonstrates to a potential research advisor, mentor, or employer that you care enough to show that you are the best candidate for the position.

2.1.2 Transcripts

Some undergraduate research programs require transcripts as part of their application, some make them optional, and others might never ask for one. But if it's required, your college transcripts should list courses taken, grades (or Pass/Fail) received, and credits awarded. For students at USA colleges and universities, official (stamped) or unofficial transcripts can be obtained from the school's Office of the Registrar.[5] Colleges and universities in many other countries do not routinely provide transcripts or may not even produce them. If a transcript is unavailable, it's a good idea to keep a list of the courses you've taken and have written documentation of how you did in them. Be open and willing to discuss your course background, what you found exciting about particular courses, or anything that is unusual on your transcript. As for the latter, every transcript is unique and we all have something unusual on them.[6] These are not barriers to success; they are opportunities to celebrate difference and diversity.

2.1.3 The Cover Letter

Although it is not the first thing we would advise you to work on in your application, the first document that a prospective mentor is likely to see is your cover letter. The goal of such a letter is to introduce yourself and convince your prospective mentor that you are a perfect fit for the position. In our experience as research mentors and readers of thousands of applications, the most compelling cover letters are those that address specific topics that show how you as an individual share the goals of the research group, organization, or institution to which you are applying (Box 2.1). If we start reading an application that begins with a generic cover letter, we rarely read the rest of the application.

Box 2.1 The essentials of a cover letter

Cover letters should be formal, short (not more than a single page), focused, and highlight and reference key points that you expand on elsewhere in your application. Five items should be included, and each should be covered in 1–2 sentences. Where appropriate, point to parts of your application that include more detail on a specific item.

1. Clearly state the research experience or position you are applying for.
2. Identify what you hope to learn or gain from this specific research experience.
3. Describe how you think this research experience and the mentor, research team, or research program can benefit from your participation.
4. Show how this research experience fits into your larger longer-term personal or professional goals.
5. Note any "red flags" on your résumé or transcript. We all have them somewhere, and acknowledging and addressing them in your cover letter demonstrates honesty and forestalls surprises during your interview.

Box 2.2 illustrates examples of effective and ineffective cover letters.

In many applications, the cover letter may be the only opportunity you have to make your case. The care you put into your cover letter is immediately apparent to anyone who reads it, and even if your application includes a separate personal statement or essay, you should still spend a lot of time and effort on your cover letter. So, after you finish your cover letter, check it carefully for typos and other errors. A surprisingly common mistake, especially when you are applying to many opportunities and using cut-and-paste functions to create multiple cover letters, is to omit or misstate the position you are applying for in the first sentence. Rather than relying on built-in spell-checkers or grammar-checkers, print it out and read it aloud to yourself or to a friend. Edit it, and read it aloud again.

Box 2.2 Effective and ineffective cover letters

Generic cover letters look something like this:

Dear Professor,
I am applying to the internship program at your University. I expect to get as much experience as I can out if this program because I know it will help in the future. This will also make my resume a lot stronger than it is now. I hope to be able to use this experience to gain employment in the future.

In contrast, a cover letter that encourages someone to read the rest of the application and learn more about you looks something like this:

Dear Professor Feynman,
I am applying to the Mathematical and Theoretical Biology Institute's summer research experience at Arizona State University. I am a Computer Science major, minoring in Mathematics, and looking to gain experience in computational methods. The program emphasizes collaboration, and I can bring to the research group skills in teamwork that I've learned as a forward on my college's rugby team. The methods and approaches I will learn at the MTBI will enhance my knowledge about career opportunities in Computer Science. I also hope to use and apply what I learn this summer in an independent research project next fall. More details on my background, skills, and qualifications for this research experience are included in my résumé, transcript, and essay.
Thank you for considering my application. I look forward to hearing from you about this exciting research opportunity.
Sincerely yours, —

2.1.4 Essays and Personal Statements

Some applications might ask for a personal statement or essay in addition to a cover letter. A cover letter differs from a personal statement or essay, and all three have different purposes. Your formal cover letter introduces you and summarizes your interest in and experiences related to the position, and highlights certain aspects of your résumé.

In contrast, your personal statement is all about you. You should think of your personal statement as an extension of the fourth point of your cover letter (Box 2.1): how the research experience fits into your broader personal or professional goals. If a cover letter is not required or expected, you might consider starting your personal statement or essay with the key details that you would have included in your cover letter. In your personal statement, you can also highlight past experiences (both good and bad) that have prepared you for the experience you're applying for. Personal statements are usually limited to 1–2 pages.

Some programs or research mentors also ask for an essay. "Essays" can serve many different purposes, and how they are written depends on how they will be used. For many research experiences, essays provide an opportunity to expand on what kinds of research you're interested in. Some mentors or programs may expect the essay to be a short proposal for a research project that you'd like to work on. Any time an essay is required, you will want to understand the motivation for, and specifics of, the essay before you start writing it. If the motivation for, or specifics of, the essay is not clear from the application form, you should ask for additional information from the program coordinator (usually accessible via a "contact us" button or form).

2.1.5 Letters of Recommendation

Although your résumé should conclude with the names and contact information for references, many undergraduate research programs will ask for letters of recommendation to be submitted with your application. Just as your cover letter, résumé, personal statement, and essay were tailored for a particular research experience or program, your references should come from individuals who know you well and can strongly support your application to a specific program. Finding these individuals and getting them the information they need to write you a strong letter of recommendation takes time, so you should line up your references well in advance of the application deadline.

Who should you ask? First, check to see if the position or program you are applying for specifies who can write your reference letter. Some programs may ask for a letter from your undergraduate advisor, someone who taught you in one of your science or mathematics courses, or a past research mentor (if you have had previous research experience). Others may not specify who can write a letter of recommendation. Either way, ask yourself, "who knows me well enough to speak about me as a person and professional?" This person could be your academic advisor, but they also could be your job supervisor, lab teaching assistant, or someone in your community with whom you have worked closely with, such as a pastor, rabbi, imam, coach, or youth-group leader.

You might be tempted to ask someone (like a professor) because they have a good title or credentials, but if they don't know you well, their letter is likely to be generic. Such generic letters are much less effective than a letter from a lab instructor that speaks to your particular talents and value to the research experience you are applying for.

We have seen recommendation letters that range from the very generic to the very specific. For example, a generic letter like this one:

> *Natasha took my sophomore organic chemistry class and received a B+, which placed her in the top 15% of the 80 students in the class. She struggled with some concepts, but came to office hours regularly for help and guidance. I was impressed with her apparent commitment and effort.*

is much less informative and is unlikely to be as helpful to you as an applicant than a more specific one like this:

> *Natasha was among the top students in the lab section for my sophomore organic chemistry class. She came up with new optimization methods that improved the class lab exercises, and has talked to me in detail about her interests in pursuing a career in organic synthesis. Her application to the HHMI undergraduate research program at MIT in AI-assisted drug development clearly reflects her career goals, and her participation in this program not only*

would accelerate her career trajectory but also would bring new ideas and perspectives to your lab group.

Because someone looking to bring you into their research group is much more likely to do so if they receive a specific letter than a generic letter, you want the person writing your letters of recommendation to know you well enough to write a very specific—and strong—letter of recommendation.

Once you have your list of potential recommenders, select the one(s) who will best support your application for a specific position. Indeed, you may want to have different people writing you letters for different applications. Make an appointment to talk to your recommender, be prepared to describe the position you're applying for, and be sure to ask two key questions: "Can you write me a **strong** letter of recommendation for this position?" and "Do you have the time to do it by the deadline?" If they answer yes to both questions, give them copies of the position or program description, your cover letter, résumé, and any other required materials. Ask for any feedback they might have on your application packet and follow up about a week before the application deadline to make sure that they're on track to submit their letter. But it's also okay if they say no to either question. Thank them for their time and move on down your list. Better to know up front that you can't get a good letter from a person and look elsewhere than to have someone whom you think will support you write a lukewarm or generic letter instead. For more on handling acceptance *and* rejection, see Sect. 2.3 later in this chapter.

2.1.6 Online Application Systems

Most undergraduate research programs use online application systems. These systems are used for several important reasons. First, online systems "level the playing field" for all the applicants by ensuring that all applications arrive in identical formats. No "extra points" are given for fancy paper or are taken away for last-minute express-mailed applications. Second, online systems allow program mentors and directors to collect aggregated data that are required by funding agencies.[7] Third, online systems provide tools to automatically remind applicants and recommenders to submit their materials on time. And finally, online systems help mentors and program coordinators rapidly screen and evaluate applications.

For all their benefits, however, online application systems can be unforgiving. As mentors and program directors, we have seen and experienced a lot of mishaps and frustration when students, recommenders, and we ourselves have dealt with online application systems. For example, online systems are automated, require forms and fields to be completed, and usually shut themselves down precisely on the day and at the time when applications are due—so be aware of requirements and deadlines (Box 2.3). Exceptions for incomplete or late applications are rarely made, and even if they could be, it may be impossible to find a person behind the machine who can help. Automated requests for transcripts, other materials, or letters of reference may end up in Spamboxes or junk folders—be sure to "allow" email or texts from application systems. And importantly, online systems require Internet access. However, lack of personal Internet access need not be a barrier; one can usually can get online at public libraries, town or city halls, or local community colleges.

Box 2.3 Pitfalls to avoid in online applications

In the world of online applications, you want your application materials to standout but you yourself do not want to be noticed. You will make a good initial impression with mentors and program leaders if you pay attention to details.

1. Whenever possible, use a computer rather than cellphone or tablet to complete online applications. Websites with online forms often work differently (or incorrectly) on hand-held devices.
2. Create and edit content in a word processor before copy-and-pasting it into online forms. Word or character limits are strict and automated systems will cut off anything beyond the limit. Be sure to check that everything you pasted into a box is actually there.
3. Special characters and formatting may not copy or paste into online forms. Whenever possible, avoid typing or pasting directly into text boxes if it is possible to upload individual files (for example, as pdfs).
4. Check content carefully in the online system to catch any errors in entering information or from copy-and-pasting.
5. Posted deadlines are real and hard. Don't wait until the last minute (or last hour) to upload and submit your application. If you encounter any technical issues, you will need plenty of time *before the deadline* to work through them. Don't expect technical support to be there for you just before a midnight deadline.

2.2 The Interview

When your efforts in crafting a compelling application are successful, you will normally be contacted for a follow-up interview. From the mentor's or program director's perspective, the purpose of the interview is to learn more about you as a whole person, to hear more than what you could communicate in your short cover letter, résumé, or personal statement, and to better gauge your interest in the research project or program. From your perspective, you want to make a solid connection with your potential mentor. You should talk about how your interests and skills align with the goals of the project and the mission (if there is one) of the overall research program. But don't talk only about your skills and education. Make a personal connection between you, the interviewer, and the research. Be clear about how this research

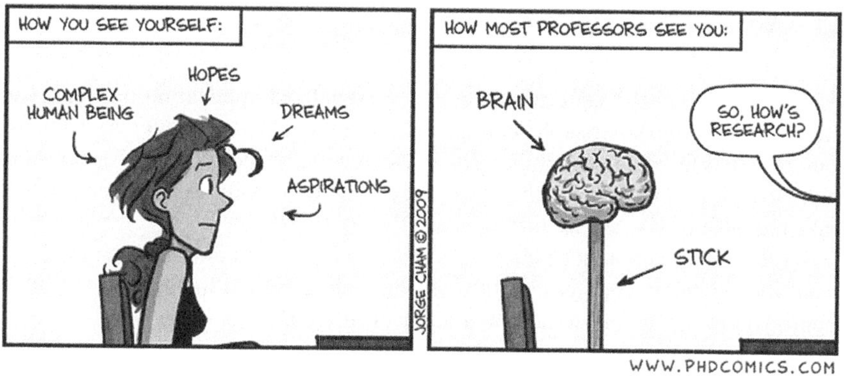

Fig. 2.1 In an interview, you want your potential research mentor to see you as a whole person, not simply as a brain on a stick. Comic strip from "Piled Higher and Deeper" by Jorge Cham https://www.phdcomics.com, and used with permission

experience would be really beneficial to you and why they would benefit from your joining the team (Fig. 2.1).

Regardless of how the interview will be conducted (phone, video, in-person), you should prepare for it. Because it may have been several months since you submitted your application, be sure to review the description of the research project or program and your application packet before the interview. Write up and be ready to ask some pointed questions during the interview. Good questions illustrate genuine interest and enthusiasm, and will lead to more questions and greater dialogue.

Finally, the setting matters. As with any interview, you should dress in a way that best represents the "professional you" while making you feel confident and comfortable. Even if no one will see you on a phone interview, there are subtle ways that your attire can influence your interview behavior (Cutts et al. 2015). If the interview is conducted over the phone or using a video-conferencing system such as Skype, Microsoft Teams, Google Meet, or Zoom, find a place that is quiet and where you won't be interrupted. Test the relevant technology beforehand. If you're using a cell phone, make sure it's fully charged and that you have a strong signal. If you're using video-conferencing software, make sure you have installed the current version with any security patches, that your camera and microphone are functioning, and that your room (or virtual background) is uncluttered and not likely to present distractions. For in-person interviews, arrive a few minutes early to familiarize yourself with the overall setting and surroundings.

2.3 Acceptance and Rejection

Ideally, you submitted your complete and personalized application on time, your references got their letters uploaded on time, too, the interview went very well, and you've been offered an opportunity for a research experience. The rest of this book goes through preparation for the experience, how it is likely to work, and how to make the most of it—before, during, *and after* it.

But it is also likely that despite all your best efforts, you didn't get an interview or an offer. Perhaps you got caught when the online system locked you out or the people you hoped would write you a letter of reference didn't get it in on time or at all. Maybe the interview was a disaster. Even if your application was complete and submitted on time, there may have been many more applicants than there were positions, and for whatever reason, your application didn't make the final cut. All of us have been rejected at one time or another, but rejection needn't end your pursuit of and participation in STEMM research.

There are four key things to remember to best move beyond rejection and find new opportunities. First, fix what you can control (timing, completeness, references, interview skills) and forget about what you can't (decisions and priorities of mentors or program directors). Second, apply for more than one opportunity, but make sure you tailor your applications for each program. Third, look for other program opportunities whose deadlines haven't passed yet. And finally, contact professors or researchers who might need part-time or summer lab help (but we don't recommend volunteer or unpaid opportunities, as they are biased towards those who can afford them and least often need additional experiences). These other opportunities can help build your résumé for a future program application. In short, persistence pays.

2.4 Take-Home Messages

- ✔ Everything about your application should talk to the position you are applying for and why you are a good fit for it.
- ✔ Respect the process and follow any specific guidelines. This shows that you pay attention.
- ✔ Rejection is a part of life but it doesn't reflect your enthusiasm or abilities.

References

Blackburn, H. (2017). The status of women in STEM in higher education: A review of the literature 2007–2017. *Science & Technology Libraries, 36*(3), 235–273.

Cutts, B., Hooley, T., & Yates, J. (2015). Graduate dress code: How undergraduates are planning to use hair, clothes and make-up to smooth their transition to the workplace. *Industry & Higher Education, 29*(4), 271–282.

Johnson, D. R. (2012). Campus racial climate perceptions and overall sense of belonging among racially diverse women in STEM majors. *Journal of College Student Development, 53*(2), 336–346.

McDevitt, A. L., Patel, M. V., & Ellison, A. M. (2016). Insights into student gains from undergraduate research using pre/post assessments. *BioScience, 66,* 1070–1078.

Seaton, G. A. (2011). Belonging uncertainty and psychological capital: An investigation of antecedents of the leaky pipeline in STEM. Master's thesis, Department of Psychology, Purdue University, Indianapolis, IN, USA.

Chapter 3
Getting Ready

Abstract Success in an undergraduate research experience requires adequate preparation. This chapter organizes the prep-work into manageable steps: (1) knowing yourself and clearly setting and communicating expectations for yourself and your research mentor; (2) dealing with the paperwork and logistics for travel, housing, food, and getting paid; and (3) reading about and understanding the research that you will be doing. In presenting and going through these steps, the emphasis is on preparing for a residential undergraduate research program, but the same steps apply for individualized internships.

In the last few months, you've searched for opportunities to do research, applied for one or more research experiences, been accepted into at least one of them, and finally, accepted an offer. Congratulations! Now, you might think that you can chill out for the next few months before you start work in your individual mentor's lab or join a larger undergraduate research program. Think again.

Your undergraduate research experience will be nothing like your science courses with their readings, lab exercises, exams, and term papers, nor is it a part-time or full-time job. Rather, it is an opportunity to do original, open-ended research in a supportive environment that also embeds you in a network of like-minded students and scholars whom you may befriend, work, and collaborate with for years to come.

Besides preparing to do the research itself, you need to prepare for the environment in which you will be working. Some undergraduate research experiences will provide housing and meals, others will not. Getting paid may require regular entry of hours worked or you might get a single stipend payment. If you have to travel from your home or home institution to the undergraduate research experience, your travel may be arranged for you, or you may have to arrange it yourself. Finally, you may find yourself outside of your personal comfort-zone while making new friends and meeting new colleagues in a new place. Something that might seem trivial to one person can be threatening or life-changing to someone else. If you expect the unexpected and plan accordingly, you won't be disappointed.

Getting ready for all of this while keeping up with your regular schedule can seem like a second full-time job. But spending some time preparing for your undergraduate

A. M. Ellison and M. V. Patel, *Success in Navigating Your Student Research Experience*, SpringerBriefs in Education, https://doi.org/10.1007/978-3-031-06641-2_3

research experience before you start it will help make it a positive, worthwhile, and successful experience.

In this chapter, we break down the prep-work into some manageable steps. These include: the value of knowing yourself and clearly setting and communicating expectations for yourself and your research mentor; dealing with the paperwork and logistics for travel, housing, food, and getting paid; and reading about and understanding the research that you will be doing. There are two Text Boxes and one Vignette in this chapter. Vignette 3.1 illustrates the importance of setting and communicating expectations. Box 3.1 presents four key ways to get ready for your research experience. Box 3.2 suggests some questions to answer before you start your research experience that, if you answer them again after you've finished it (see Chap. 7), will allow you to see the real progress you've made in learning about and doing STEMM research. In presenting and going through these steps, we focus on preparing for a residential undergraduate research program, but the same steps apply for individualized internships.

3.1 First, Know Yourself

To undertake and participate effectively in a research experience, you need to be healthy, both physically and mentally. In Box 1.1 in Chap. 1, we encouraged you to reflect on personal aspects that would put you in a mental or physical space to be able to fully embrace and engage with a research experience. For example, would it work better for you if you were close to home or further away? What kind of cultural environment or living situation would enable you to thrive (Vignette 3.1)? Do you have other responsibilities or obligations that might make it difficult to handle the logistics of doing the research or meeting the expectations of your mentor?

Everyone doing research (or any other internship or job, for that matter) has a life outside their work, and one of the more remarkable aspects of the research enterprise is the flexibility and willingness of most everyone involved to help each of us "make it work." But to take full advantage of that flexibility, and to become aware of limits to it, you need to know yourself and your own needs, be able to clearly and honestly communicate and discuss them to your research mentor or research program staff, and be willing to compromise when you can. In any research experience worth undertaking, it will be fine to talk about your individual needs and ask for necessary accommodations, but you need to take the first step: speak up!

Vignette 3.1 Understanding and communicating your needs

Based on the experiences of the authors

Alex applied for and has been offered via email a position in an undergraduate research program with a residential component. During the interview, Alex asked about the option of having a single room (this was an expectation that Alex had for joining the program). The interviewer (in this case, the mentor) stated that a single room would be a possibility and room assignments would be made based on a housing survey to be completed by the student about one week before the program started.

Meeting Alex's expectation in the context of the program's logistics was accomplished through an email exchange:

February 15
Dear Alex,
We'd like to offer you a position in our undergraduate research program. …If you have any questions, do not hesitate to let me know.
Best,
Jane
Program mentor
she/her

February 16
Dear Jane,
I am happy to accept the position! I do have one question right off the bat: Will the program be able to accommodate my request for a single room that we discussed during the interview?
Thanks,
Alex
they/their

February 17
Dear Alex,
We are excited that you will join the program! In terms of your request, yes we are able to assign you a single room. If you have any other questions, do not hesitate to contact us.
Best,
Jane
she/her

Two months later …

May 1
Dear Jane,
During my interview and acceptance, I was assured that I would be able to get a single room as that would be the most comfortable option for me. I just wanted to check in and confirm this is still possible.

Thanks,
Alex
they/their

May 2
Dear Alex,
Thank you for the follow-up email. Yes, we have your choice of a single room noted and will assign one to you. Next week, we will be emailing the housing survey to all students. We base all our assignments on student responses. Rest assured that you will have a single room.
Best,
Jane
she/her

May 3
Dear Jane,
Thank you very much! I'll fill out the housing survey as soon as I get it. I'm very much looking forward to joining the program in a couple of weeks.
Best,
Alex
they/their

3.2 Pay Attention to the Logistics

Whether your undergraduate research experience is on your home campus, at another institution, or in another city or country, there will be new policies, procedures, and protocols to pay attention to. There may be forms to fill out, on-line training courses to complete (see Chap. 5), and new lines of communication to establish. For many of these, you'll be working with your research mentor's assistant or a program coordinator. Either way, working and communicating well with them will smooth the path to a successful undergraduate research experience.

In our experience, students who put time and effort into getting the logistics done well and on time are individuals who become successful researchers. Conversely, students who slack off on the logistics or fail to communicate effectively and in a timely manner with program staff also tend to communicate ineffectively or not at all with their mentors, be poorly prepared for their research, and sloppy in the lab or field. Here are three tips to make the time and effort spent on logistics planning pay off:

- **Read.** If you received an acceptance letter or package, read it thoroughly. Highlight the key points (if it came by email, print it out) and put them in your to-do calendar.

It's best to make the "do-by" date a few days before the deadline. To quote one of our former student researchers, *on-time is late.*

- **Consider.** Remember that program coordinators, administrative staff, and your research mentor are doing other things at the same time as they're preparing for your undergraduate research experience. If they ask for some seemingly trivial logistical task to be done at a certain time, there's probably a good reason for it. Don't hesitate to ask why, but once you get the answer, follow up and follow through promptly.
- **Learn.** Understand how your research mentor, program coordinators, and administrators want to communicate. Be prepared to adapt to them (after all, it's their lab or program and they're paying you). If they prefer to use email, you should, too. If you would prefer to use text, Slack, or some other messaging system, ask if it will work for your mentor or the program. You may find yourself using different modes of communication with different people and for different parts of your undergraduate research experience.

3.3 Familiarize Yourself with the Research

Although it's probably been a few months since you submitted your application, remember that you applied to an undergraduate research experience for one or more reasons (if you don't remember those reasons now, go back and review your answers to the questions in Box 1.1 in Chap. 1). It's likely that one of the more important ones was the type of research you'll get to do. The project description was probably pretty short and you learned only a little bit more about it when you interviewed for the position. Because undergraduate research experiences may last only a couple of months or a single semester, being ready to dive into the research on day one lets you get the most out of it. So, before you show up to start your research experience, spend some time learning more about the research project itself and the broader context in which your soon-to-be research mentor is working (Box 3.1).

Box 3.1 Preparing to do your research

Here four key ways to get ready for your research experience:

1. **Review any materials**, including background readings, research articles, or protocols, that were highlighted in the description of the project you applied for, discussed during the interview, or sent to you along with your acceptance letter or shortly thereafter.
2. **Get a sense of the research environment** you're about to join. Look over the websites of the program and your research mentor. Even if you did this while you were applying, reviewing it again may reveal

things you overlooked before. Many programs, research mentors, and students who were previous participants in the same program or research group have blogs, vlogs, YouTube channels, or are active on social media. Read, watch, friend, or follow them.

3. **Read some recent articles by your mentor**. Most researchers will have links to pdfs of their articles on their lab or personal websites. Although reading primary research literature can be a bit daunting if you haven't done it before (or even if you have), don't be overwhelmed or intimidated. Remember that your goal here is to get a sense of the big picture of your mentor's research. When you come across parts that are difficult to understand, write down questions to discuss later with your mentor.
4. **Email any questions** you may have to your research mentor before you start. If you think that learning the answers now will help you be better prepared to do the research, say so. Your mentors want you to have a successful undergraduate research experience, and they will be happy to provide answers and guidance. The best way they can help you prepare for the research is by understanding what you don't know and would like to learn.

3.4 Prepare for the Imperfect

An undergraduate research experience, just like a career in research, challenges us to push the boundaries of knowledge and understanding. But doing research also involves applying skills—oft-times repetitively—to particular tasks that may be challenging and fun or tedious and dull (Chaps. 4–6). Many aspects of your undergraduate research experience will be easy to do and fun to accomplish; it is natural that you will want to spend more of your time on those parts and less on the parts that are difficult and require more time, practice, attention, or repetition. These latter parts could be specific tasks such as pipetting serial dilutions, rearing parasitic fungi from caterpillars, running statistical tests, or coding a simulation model. Alternatively, you may not enjoy the interpersonal aspects of networking with your peers or building a rewarding professional relationship with your mentor, especially if you're not "clicking" with your fellow students or your mentor for whatever reason. Overcoming these challenges can add value to your undergraduate research experience and shape your future relationships with your own mentees (see Chap. 9).

These challenges can also impact your physical and mental health during and after your undergraduate research experience. Before you start your research experience and when you are calm and safe, take some time to think about what you could do if things go wrong. Think back to a time in a class or at work when you expected something to go a certain way and it didn't. A deadline got moved up, you got scheduled to work the night before finals, or you got a bad grade on an exam or paper. What did you do? As with most of us, you probably got sad or angry. But you also moved on and forward, learning from the experience. Were there things you could have changed or done differently? Would more open communication have helped? If you can anticipate similar situations or contexts arising during your undergraduate research experience, talk to your mentor or program staff in advance and manage your—and their—expectations.

3.5 What Will a Successful Experience Look Like to You?

How will you know if your undergraduate research experience was a success? What criteria would you use to decide? If you could come up with a set of questions that you would use to define a successful research experience, what might your list include? Would they include questions whose answers are descriptive and qualitative or numeric and quantitative?

Research mentors, undergraduate research program directors, and funding agencies have expectations for successful research experiences. Based on these expectations, these individuals and organizations have developed surveys or questionnaires to get information from you and other student researchers. But you might not have thought to ask these same questions of yourself.

In the spirit of using a research experience to better know yourself, we suggest you ask yourself some questions about your own skills and measures of success (Box 3.2).[1] If you pose these questions to yourself now, and then revisit them at the end of your research experience (see Chap. 7), you will probably surprise yourself with how much a research experience changes you.

Box 3.2 Some questions to ask yourself before you start your undergraduate research experience

1. Am I looking forward to my research experience?
2. Have I been mentored in research before?
3. Will I enjoy working on a research team?
4. Have I ever felt like a respected member of a scientific research team?
5. How well prepared am I to collect, analyze, write up, or present scientific data and results?
6. What are my short-term (1–3 years) goals for employment or additional education?
7. What are my long-term (5–10 years) goals for employment or additional education?

8. How likely is it that either my short- or long-term goals involve research in a STEMM field?
9. How confident am I in my answers to the previous questions?

Feel free to add more questions of your own!

3.6 Take-Home Messages

✔ Prepare for your experience by learning about the context and methods of the research you will do.

✔ Logistics matter. Read program materials thoroughly, and follow through by responding to everything you are asked for.

✔ Know yourself and pro-actively identify potential challenges so you can get the most out of your undergraduate research experience.

Part II
Undertaking Your Research

The three chapters in the central part of this book consider the undergraduate research experience as a "three-legged stool," whose legs are research, education, and community. First among equals, an undergraduate research experience gets you doing real research. Second, an undergraduate research experience provides you with opportunities to learn about the broader context of a research topic, grow your own knowledge, and learn new skills. And third, becoming a researcher introduces you to, and embeds you in, the large and supportive community of scientific researchers. The three legs are equally important and the balance among them will guide your future in STEMM.

Chapter 4
Research

Abstract The undergraduate research experience can be conceptualized as a "three-legged stool," whose legs are research, education, and community. First among equals, an undergraduate research experience gets you doing real research. In the context of an undergraduate research experience, "research" should be viewed simultaneously as an opportunity to do research, a learning experience, and a job. This chapter continues the discussions begun in Chap. 3 about personal goals and expectations, and delves further into understanding and meeting expectations. The importance of communication with your research mentor is emphasized, so guidance and best practices for successful mentee-mentor communication is provided. This chapter also introduces different measures of "success" in research and illustrates why one's sense of self-worth and success in research should not be tied to statistical "significance."

Research is the first leg of the three-legged stool that is the undergraduate research experience. In this chapter, we focus on this component, which should be viewed simultaneously as an opportunity to do research, a learning experience, and a job. We continue the discussion that we began in Chap. 3 about personal goals and expectations, and delve further into understanding and meeting expectations. We emphasize the importance of communication and provide some guidance and best practices for successful mentee-mentor communication. The one Vignette in this chapter (Vignette 4.1) discusses why you should not tie your sense of self-worth and success in your research experience to statistical "significance."

4.1 Doing Research During an Undergraduate Research Experience

In your undergraduate research experience, you should expect to really do research. As we discussed at the beginning of Chap. 1, research starts with asking a question that you are interested in *but don't know the answer to*. Therefore, the first step in

A. M. Ellison and M. V. Patel, *Success in Navigating Your Student Research Experience*, SpringerBriefs in Education, https://doi.org/10.1007/978-3-031-06641-2_4

doing undergraduate research is for you and your mentor to work together to identify an interesting question.

Your mentor may already have some idea of how to best approach answering the question, but in most cases you and your mentor will work together to determine the necessary methods, tools, and techniques you will need to use to answer it. Alternatively, your mentor may ask you to figure them out on your own. Next, you'll use the methods to make observations and do experiments, and collect and analyze the data. Finally, you'll circle back to see how your results help answer your question, and what new questions they suggest. Keep in mind, though, that the process of doing research, identifying new questions, and opening up new avenues for future research is often more important than getting "the" answer.

But there's more to doing research than asking questions and developing and using methods to answer them. Any research experience—indeed, any time spent doing research—is both an unparalleled learning opportunity and a job. Successful researchers approach doing research as a combination of the two. Both jobs and learning opportunities have clear and unspoken expectations you will need to meet. If you can get the job part right, you will have plenty of time for learning and doing.

4.1.1 The Job of Doing Research

What are the basic expectations for a job? Jobs usually have set working hours—they are called "9–5" jobs for a reason. Research positions are more often defined by the question and work involved in addressing it and usually have some flexibility in working hours. But whether you work in a lab or an office, in the field or online, there's likely a minimum number of hours for which you are expected to work. If your research experience requires you to be on site in the lab or in the field, or online for a certain set of hours on each of a fixed set of days, your mentor should have let you know that during the interview (see Sect. 2.2 in Chap. 2) and in the letter offering you the position (see Sect. 3.2 in Chap. 3). If you have some flexibility in defining or setting your "normal" work hours, it's a "best practice" to actually define your hours, let your mentor and any other members of the research group know what they will be, and stick to them.

As part of your research experience, you may be assigned to learn and complete certain tasks at regular intervals (e.g., hourly, daily, weekly, monthly). Such tasks might include preparing standards or reagents for chemical procedures or analysis, counting cells or bacteria under a microscope, measuring trees, scanning the skies for comets, or entering data into a spreadsheet. Most research projects involve collecting a lot of replicated data, and obtaining these replicated data usually means doing the same thing many, many, many times. As with any job, you will learn to do certain tasks in a certain way and your skills will improve with time and practice.

But at the same time, you should be thinking about these tasks and observing their outcomes. For example, your mentor may have asked you to count bacterial colonies on several hundred cell-culture plates. As an employee doing a job, you would set about doing the task: you'd count the bacteria in an eight-hour work day and then head for home. But as a researcher, you might also notice how the colonies grow, their patterns on the plates, and the interactions between them. You might try to characterize the patterns and ask why you observe only some patterns and not others. In short, you're not just counting colonies, you're asking "what is going on here"? And you may not stop wondering about it when you leave the lab at the end of the day.

4.1.2 Learning While Doing Research

So, what are you learning while doing the repetitive tasks required to collect the data you need? Doing research provides a physical space and opens a mental space where it's safe and expected for you to ask questions and further your and others' understanding of science.

There are two ways to grow your understanding: learning from others (and especially from your mentor), and learning by and from yourself. During your undergraduate research experience, your mentor should take the time to help you understand the broader context and the specifics of the research you are doing.[1] This help can come during formal or informal meetings and discussions, by suggesting or providing you with journal articles (usually) or monographs and textbooks (less commonly) to read, or by giving you the time and space to explore questions and ideas on your own and then giving you feedback on your explorations. Don't be shy! Your research experience will be richer if you continually ask yourself and your mentor more questions, spin more ideas, and generate plausible hypotheses. Remember that it's research: when you're pushing the boundaries of knowledge, there are no stupid questions and no wrong answers.

4.2 Articulate Your Goals

Both you and your mentor should have clear goals for a successful research experience. Your mentor and any departmental or program staff will expect you to be a careful, aware researcher and to participate fully in available opportunities for additional education and becoming part of the research community. But you should also have goals for yourself, your mentor, and, if you are in one, the undergraduate research program. It may be difficult for you to set these goals, especially if you haven't done research before or think you have less knowledge or fewer skills than others in the lab or the program.

Remember that everyone brings their own background, knowledge, strengths, and skills to an undergraduate research experience; that research mentors and other senior scientists were students—and novice student researchers—themselves; and that the most important goal of a mentee-mentor relationship is for the more experienced person (mentor) to guide the less experienced one (mentee) to new levels of understanding (National Academies of Sciences, Engineering, and Medicine 2017, 2019).[2] In setting and articulating your own goals for your undergraduate research experience, you are showing that you are serious about the research experience and want to have meaningful gains from it.

It is important to always remember that this is *your* undergraduate research experience. If you take an active role in defining the experience and framing your goals and expectations—for yourself, your research mentor, and the research program—you will be much more likely to have a successful experience. Communicating your goals, especially to your research mentor, is vital. Starting this communication as soon as you've been accepted into a lab or program is better than waiting until the day you arrive (see Chap. 3). Different methods of communication—phone, email, text, video-chat, in-person—are appropriate for different situations, and some are considered more formal than others. Think about how you would communicate with your friends, your teachers, or your employers; context matters in an undergraduate research experience, too.

4.3 Setting and Meeting Expectations

All professional practicing scientists were students once, had mentors who helped them learn about the practice of science, and probably still have mentors from whom they are still learning. Few practicing scientists ever work alone or in isolation, and as a student researcher, you need peers, colleagues, and mentors, and co-learners working with you side-by-side. Although you do need help and guidance from others with more time and experience in the field, you also want to push yourself into learning and understanding something new. Setting goals with clearly defined expectations is a good way to push yourself and make it more likely that you will have a fulfilling and successful research experience. How can you best set your own goals and expectations? And once you've set them, how do you go about meeting them?

4.3.1 Communicate Clearly and Openly

The process of meeting goals and expectations starts with articulating your own goals and expectations. Do you want to learn a particular skill, develop an idea for an independent project or senior thesis, attend a meeting and present your research to your colleagues, or publish a scientific article? Do you want to meet other student researchers, learn about opportunities for jobs or advanced study (e.g., graduate school) in STEMM fields, or improve the scientific literacy you will bring to other parts of your life or a future career? Before continuing on, take a break from reading and make a list of your expectations and goals, and prioritize them.

My top five goals and expectations as a student researcher, ordered from highest to lowest priority

1.
2.
3.
4.
5.

Keep in mind that these goals and expectations are not fixed in stone. They will evolve and change as you progress through and beyond your research experience. It's a good idea to revisit them on a regular basis.

At the same time, your mentor will have goals and expectations for and from you, and these may not align perfectly with yours.[3] Discuss your list with your mentor(s), and *vice-versa*. Your mentor(s) will listen, provide feedback, and hear your responses. At the end of what will be the first of many conversations, you'll find some middle ground that works for everyone. Research is like life: you can't always get what you want, but by being open and flexible, you can get out of your research experience what you really need (and maybe some unexpected things too). At the same time, there probably are several ways to reach the same goal.

4.4 Measuring Success in Your Research

> In 1953, a fledgling company called Rocket Chemical Company and its staff of three set out to create a line of rust-prevention solvents and degreasers for use in the aerospace industry. Working in a small lab in San Diego, California, it took them 40 attempts to get the water displacing formula worked out. But they must have been really good, because the original secret formula for WD-40® Multi-Use Product—which stands for Water Displacement perfected on the 40th try—is still in use today.
>
> — From the history of WD-40

Imagine if they had stopped after the 12th try. But they didn't, and WD-40® is a world-wide best-seller. Research is really a process: asking an interesting question, getting *an* answer, and identifying new questions along the way. There's rarely only one answer to a research question, and even not getting an answer is still an answer. All scientists ask questions that they can't answer and pose hypotheses that lead to dead ends or turn out to be flat wrong, but we all learn something along the way. Learning continuously and building on experience—both successes *and* failures (see Vignette 4.1)—is at the core of successful scientific research. And by the time you've finished your research experience—and worked your way through this book—you'll have many other ways to define and measure your success, too.

Vignette 4.1 "Significance:" What does it mean?

Contributed by Harrison Mancke

When we hear the words "significant results" we typically expect those results to support our hypothesis. It's an odd bias that young scientists have to work on dismantling early in their career. In 2019–2020, I was still very new to science research. I studied acute copper exposure in fish to test for sub-lethal effects based on the EPA's water quality criterion (WQC). My study found "no significant difference." I was quite bummed about it. No significant difference? My study was useless!

Although my study showed no significant differences between the control and experimental groups, that didn't mean my results were insignificant or useless. Rather, they showed that the current EPA WQC was, in fact, suitable to minimize sub-lethal effects.

The only situation I can imagine where a scientist would have "insignificant results" is if their study just didn't work (e.g., the lab caught on fire, all the samples were stolen, etc.) You can learn nearly anything from your data, as long as it is within the realms of reality. To that end, a significant result may not be what you're expecting—so keep your mind open, and try to diminish bias.

4.5 Take-Home Messages

✔ Research is a process that continually leads to answers, non-answers, and more questions.

✔ Articulate your own expectations and goals for your undergraduate research experience.

✔ Success comes in many forms, so don't get caught up in only one of them.

References

National Academies of Sciences, Engineering, and Medicine. (2017). Undergraduate research experiences for STEM students: Successes, challenges, and opportunities. Washington, DC: The National Academies Press.

National Academies of Sciences, Engineering, and Medicine (2019) The science of effective mentorship in STEMM. Washington, DC: The National Academies Press.

Chapter 5
Education

Abstract Education is the second leg of the three-legged stool that is the undergraduate research experience. Researchers are always learning new tools and techniques that help do research better, but there are a wide range of other skills that are needed to learn and use regularly to advance a STEMM career. This chapter discusses many of these other important professional skills. Some of these skills may be readily apparent, such as how to write an abstract, a paper, or a research proposal, and how to prepare and deliver an oral talk or poster. Others are necessary but are used "behind the scenes," and include how to effectively document and archive data and discover data archived by others, build a professional network, or explore career options and take the next steps in a STEMM career. Throughout this chapter, there is an emphasis on responsible conduct of research: ensuring that research conforms to ethical norms and guidelines, and applicable regulations and laws.

Education is the second leg of the three-legged stool that is the undergraduate research experience. While doing our research, we are always learning new tools and techniques that helps us do it better (see Chap. 4). But there are a wide range of other skills that you will need to learn and use regularly to advance your STEMM career. There is one Text Box and one Vignette in this chapter. Box 5.1 provides a checklist of important professional skills. Some of these skills may be readily apparent, such as how to write an abstract, a paper, or a research proposal, and how to prepare and deliver an oral talk or poster. Others are necessary but are used "behind the scenes," and include how to effectively document and archive your own data and discover data archived by others, build your professional network, or explore career options and take the next steps in your STEMM career. All successful scientists should learn how to ensure their research conforms to ethical norms and guidelines, and applicable regulations and laws.

A. M. Ellison and M. V. Patel, *Success in Navigating Your Student Research Experience*, SpringerBriefs in Education, https://doi.org/10.1007/978-3-031-06641-2_5

Box 5.1 A checklist of professional skills in STEMM

This list includes what we think are the most important non-technical professional or "survival" skills in which any research scientist needs to be proficient. They are listed in approximate order of decreasing importance to undergraduate researchers, but it's never too early to learn skills, such as team management, that you might be able to use later in a STEMM career.

- Doing research responsibly, ethically, inclusively, and safely
- Efficiently searching and assimilating the primary literature
- Using software tools, including word-processors, spreadsheets, and analytical software
- Preparing and presenting oral talks and posters
- Writing scientific abstracts, technical papers, and grant proposals
- Preparing effective applications for post-graduate study, employment, fellowships, and grant proposals
- Developing and managing research budgets
- Purchasing and maintaining inventories of research supplies
- Working collaboratively and managing a research group

In the past, up-and-coming scientists learned these professional skills indirectly by watching and imitating their mentors, failing repeatedly, and improving through hard experience. We now know that there are better, more structured ways to learn and develop what used to be called "soft skills." These are now recognized as being "survival skills" that are as crucial to success in STEMM as is expertise with lab, field, measurement, and analytical techniques. Most larger undergraduate research programs offer training sessions and hands-on workshops in some or all of these. Vignette 5.1 discusses how valuable such workshops can be for undergraduates and other early-career researchers. But even if you are the only undergraduate in your research group, there should be plenty of opportunities available to you to learn these skills on your home campus or from a variety of online sources. And regardless of whether you are a solo undergraduate researcher or part of a larger undergraduate research program, you should always be working to build and expand your knowledge base.

5.1 Training

Any scientist—those just starting out or those who have been working in the field for decades, and studying or working in the nonprofit or for-profit public or private sectors, in education or industry, or in government or with non-governmental organizations—need to take regular, refresher, or one-time training courses that inform us about and keep us up-to-date with required policies and protocols. As a student, you may already take certain training courses as part of your classes, to qualify for and keep a job, or to raise your general awareness. The STEMM realm is no different. There are training courses that researchers around the world need to take to ensure ethical and responsible conduct of research, diversity and inclusivity, lab and field safety, and many others that apply to specific fields of STEMM research and education.

5.1.1 Ethical and Responsible Conduct of Research

The responsible conduct of research ("RCR") is the most essential part of the scientific enterprise. But what is RCR, why is it important, and how do you go about learning it?

RCR is the umbrella term for promoting the goals and objectives of scientific inquiry, and creating and nurturing a collegial and inclusive environment that allows diverse scientists to work together toward agreed-upon, common goals. Practiced regularly and seriously, RCR results in scientific data, results, and outcomes that are trustworthy.[1] We used to learn about RCR by hearing whispers about irresponsible, unethical, or illegal conduct, such as data fabrication or falsification, plagiarism, financial mismanagement of grant funds, or harassment, and being urged not to do such things. Occasionally, unethical conduct by scientists makes headline news. But whether irresponsible and unethical conduct of research happens *sotto voce* or in your face, it can and does lead to censure, retraction of papers, suspension of grant funding, loss of employment, or criminal prosecution and conviction. It is now routine and expected that scientists at every career stage, including undergraduates, take RCR training sessions or workshops before they start to do research. Scientists are also expected to take refresher training courses in RCR at regular intervals (at least every year or two) to stay up to date.

RCR workshops emphasize positive behaviors while illustrating negative behaviors. Common goals for RCR include:[2]

- Developing and nurturing a culture of integrity;
- Empowering researchers to hold themselves and others to high ethical standards while discouraging and preventing unethical conduct;
- Increasing knowledge of, and sensitivity to, ethical issues in the conduct of research by scientists from diverse backgrounds and cultures;
- Increasing the appreciation for what are acceptable and ethical scientific practices, and what are the regulations, policies, statutes, and guidelines governing the conduct of research in one's home institution or country and in other institutions or countries;

- Improving the ability to make responsible choices when faced with ethical dilemmas in research practice.

The subject matter of RCR workshops begins with general topics applicable to all STEMM fields. These overview workshops usually take several hours and include topics such as:

- Responsibilities and relationships between mentors and mentees;
- Civil behavior, including harassment, bullying, and other inappropriate behaviors;[3]
- Definitions and examples of research misconduct and questionable research practices;
- Data management, including data collection, keeping good, written (electronic) records, data ownership, data archiving, and data sharing;[4]
- Issues around publications (McNutt et al. 2018), including guidelines for (co)-authorship, peer-review, copyright, and open-access;[5]
- Identification of conflicts-of-interest, which can include financial conflicts and interpersonal conflicts;
- Overviews of issues regarding safe operations in the lab or field and doing research on human or animal subjects;
- The relationships between scientists, the environment, and broader societies.

5.1.2 *Specific Training Courses*

Regardless of your specific field, you can expect to take additional training courses on some of these topics, such as lab or field safety, sexual harassment, or civil behavior. Depending on your research project, you may be required to take other training courses and workshops before you start your research. These might include training courses on doing research with vertebrate animals or human subjects, both of which are very tightly regulated everywhere in the world.

Don't neglect these RCR and other training sessions. Many are required before you can start your research work. For example, you can't walk into a lab until you've taken your basic lab safety course and you can't start work in the lab until you've learned about how to handle hazardous materials.

If your research involves observing or manipulating animals you can't start your lab or field work, or collect your animals until you've taken a course in research involving animals *and* received approval for your research from an appropriate oversight committee or board. In most countries, only research on vertebrate animals is regulated, but this is evolving to include invertebrates, too. Ferdowsian and Beck (2011) provides a general overview of the evolution of ethics involving research with animals, and Drinkwater et al. (2019) discusses ethics involving research with invertebrates.[6]

Similarly, if your research uses people as research subjects, which includes (but is not limited to) using tissue or genetic samples collected from people or conducting surveys or psychology experiments, you can't start your research or even distribute surveys or enroll participants in your studies until you've completed a required training course on research involving human subjects *and* received approval for your research from an Institutional Review Board.[7]

Training sessions for most of these topics are offered regularly on every college or university campus, and many are available online. Your mentor or a departmental administrator should be able to tell you which ones you are required to take before you start your research, which can wait until you're already working, and how and where to sign up. But if they haven't given you this information before you're supposed to start your research experience, you should ask. That way, you can take full advantage of the time you'll have available to do your research.

5.2 Workshops to Build a Skill-Set to Grow Your Career

Learning skills for professional development happens in the same way as learning any other skill. In STEMM, you are most likely to learn something new when you need to get a particular task done. You might learn to use a balance when you need to weigh samples, a particular statistical method when you need to analyze a specific dataset, or a microscope when you are doing dissections or microsurgical techniques. You may have initially learned each of these skills because of a pressing need, but they also can be used again, applied in different circumstances, and taught to others. Throughout your career, you will keep learning new technical skills and modifying the ones you already know to meet new needs.

In parallel, you can learn career-building professional skills when you need them. You are most likely to be open to learning about how to write a résumé or *curriculum vitae* when you're applying for a research experience, an internship, or a job (see Chap. 2). Similarly, you will be motivated to learn how to write a research proposal when you need to apply for a grant to support your research, how to write an abstract when you're ready to present your work at a conference or submit a paper to a journal, and so on. Just like weighing samples, using a telescope, or analyzing data, professional development skills can be learned, practiced, and taught to others.

These skills are often taught in hands-on workshops, and learned and honed through practice. Unlike training courses, which are focused on behavior modification and compliance, skill-building workshops emphasize thought and creativity, and act as a bridge between learning concepts and methods in a classroom setting and their real-world applications. You usually have to pass a test to complete a training course, but workshops aren't graded. Rather, you get to practice and receive constructive feedback on your work.

Lab groups and journal clubs, where you and others in your research group get together to read and discuss new articles can function as workshops in learning how to read and understand the primary scientific literature. Writing workshops or hackathons focused on a particular topic or problem provide opportunities to expand your writing or coding skills in a context that you are likely to care about.

If you're the only undergraduate or one of just a few undergraduates in your research group, talk to your mentor or other members of the group about how you might find workshops to learn professional skills such as those listed in Box 5.1. You may find that they hadn't thought before about the need for such skill development and you may spark their interest in organizing a workshop. There may be workshops available in places you (or they) hadn't considered: college writing centers may have workshops on technical writing, career centers may offer sessions on putting together résumés or *cvs*, and the statistics or mathematics department may offer individual consulting or short courses on data analysis or hold regular hackathons. There are also tutorials posted online on YouTube and more formal workshops offered online by non-profit and for-profit organizations.

Most larger undergraduate research programs offer workshops in professional skills that are tailored for the specific needs of undergraduates. Sometimes they are a required part of the program, but even if they are "optional," you should take advantage of the opportunities. In most cases, these workshops will present skills and materials that you can use effectively to advance your career in STEMM, but unlike technical skills, they are not as likely to be taught in your regular science classes.

Vignette 5.1 Building professional skills through workshops

Contributed by Julia Barefoot

Participating in workshops shows a desire to expand one's knowledge and build a variety of valuable professional skills. The opportunity to advance your skills and receive criticism and positive feedback from unbiased sources allows you to communicate research effectively with others. That's exactly what workshops can give you. In my experience, professional workshops allow you to gain perspective, especially in research, where you spend so much time focusing on a typically miniscule topic. I have had the chance to look outside the factual information my research covers by listening to other researchers from different backgrounds and with different technical skills. This allowed me to focus in on the way research is presented and how I can convey a specific emotion through conversation. I feel that professional workshops also allow you to network with individuals you otherwise wouldn't have met. As a young researcher, being able to use virtual platforms during workshops has connected me to mentors who may be multiple states away but have valuable technical and personal experiences that have contributed greatly to the future success of my research and my continued ability to learn.

5.3 Research Seminars

As you walk down the hall to your lab or browse the department website, you're likely to see notices for research seminars. What are they? Research seminars ("seminars" for short) are (usually) 30–45-min presentations by local or visiting researchers on their most recent, and often unpublished, work. On most college or university campuses, on any day during the academic year, you are likely to be able to find a seminar in some department. Seminars are an excellent and focused way for you to learn and discuss current research trends and see the big picture of a STEMM field while learning more details and gain depth in your own area of research, or being introduced to new ideas outside of your research area. At research seminars, you will meet and engage with other students and researchers at all career stages, any of whom could become part of your professional network.

Summer undergraduate research programs often invite several researchers to present seminars on their work to all the students. Even if the seminar topic is not exactly what you are interested in, you should still plan to go to the seminar, listen, and ask questions. And you never know what will stick from a seminar. If it seemed useless at the time and you were forced to sit through it, some years later you may remember it and find value in it. Similarly, if the topic was really dull or the seminar was delivered by a speaker as dynamic as a cardboard cut-out, you will have seen and learned how not to give a talk and how to do better when the time comes for you to give a seminar on your research.

5.4 Take-Home Messages

✔ Learn and understand how to ensure you conduct your research responsibly and ethically.
✔ Build your professional skill set through training, workshops, and seminars.
✔ Broaden your scientific horizons by attending research seminars in a range of topics.

References

Drinkwater, E., Robinson, E. J. H., & Hart, A. G. (2019). Keeping invertebrate research ethical in a landscape of shifting public opinion. *Methods in Ecology and Evolution, 10*(8), 1265–1273.

Ferdowsian, H. R., & Beck, N. (2011). Ethical and scientific considerations regarding animal testing and research. *PLoS ONE, 6*(9), e24059.

McNutt, M. K., Bradford, M., Drazen, J. M., Hanson, B., Howard, B., Jamieson, K. H., et al. (2018). Transparency in authors' contributions and responsibilities to promote integrity in scientific publication. *Proceedings of the National Academy of Sciences, 115*(11), 2557–2560.

Chapter 6
Community

Abstract This chapter discusses the third leg of the undergraduate research experience: the importance of being part of a community of professionals. Because scientific research increasingly is done in collaborative teams, the focus here is on learning and practicing team-building skills and on the importance of having systems of support when research experiences become intense or overwhelming. There is great value in using a research experience as an opportunity to learn from many others, including peers, mentors, and support staff.

Science is rarely a solitary activity, and your community of researchers is as important as research and education in supporting your STEMM career (Wuchty et al. 2007).[1] In this chapter we address the importance of communities of practice in successful undergraduate research experiences. Because scientific research increasingly is done in collaborative teams, we discuss how to learn and practice team-building skills and the importance of having systems of support when research experiences become intense or overwhelming. We emphasize the value of using your research experience as an opportunity to learn from many others, including peers, mentors, and support staff. There is one Text Box and one Vignette in this chapter. Box 6.1 lists the most important elements of successful research teams. Vignette 6.1 illustrates the importance of a supportive network for learning and doing research.

6.1 What are STEMM Communities?

As a person starting out in research, who should you include in your professional community? And is there only one, or will you be a part of multiple, overlapping or non-overlapping communities? For example, your primary professional community could be the other undergraduate students in your research group, together with the lead researcher and any research assistants, graduate students, or post-docs in the same group. If you are in a structured undergraduate research program, you will also be part of a cohort of undergraduate researchers, a group of research mentors,

A. M. Ellison and M. V. Patel, *Success in Navigating Your Student Research Experience*, SpringerBriefs in Education, https://doi.org/10.1007/978-3-031-06641-2_6

the program director, and the support staff who keep the whole program working smoothly. And as you advance in your STEMM career, your networks will expand to include other researchers who share your interests (such as in professional scientific societies) or with whom you routinely collaborate and co-author papers or research proposals. STEMM communities can be local, regional, national, or international, and many of their members may interact only virtually and never meet in person.

Why is being part of a professional community so important for you as a new researcher? A professional community with a common purpose instills a sense of belonging in its members, applauds their successes, and encourages them when they are less successful. When you become part of a professional STEMM community, you may get more excited by the prospect of becoming a scientist. This can be especially important if earlier in your life you were discouraged from "doing science." Like any network or community, STEMM communities can be open, warm, and welcoming or closed, cold, and insular. Seek out the former and avoid the latter.

6.2 Your Communities are Your Support Systems

The life of a researcher, like life in general, has its ups and downs. There's nothing like the high you'll get when your observations make sense, your experiments "work," your papers get published, and your grants get funded. But when you're struggling or "failing," the lows can be really deep and devastating. Having a system of support is essential for success in research, whether it's during an undergraduate research experience or much later in your career. But as a new researcher, these systems of support are essential to your success, health, and well-being. We emphasize that a research team or network is a better system of a support than a single mentor or friend. One person can't be everything for you any time you need them. A network or community of peers, mentors, coaches, motivators, and emphathizers will be better able to support you when you need it.

6.3 Team-Building

Traditional course work is done in a mostly independent format. You go to class, you study, you take the test or write a paper, and then you get a grade. But the best and most impactful research in STEMM, and also in the humanities and social sciences, is done in collaborative teams (Guimerà et al. 2005; Institute of Medicine 2007; Wuchty et al. 2007).[2] Thus, learning and understanding how to work in teams is a crucial survival skill for success in STEMM (see Box 5.1 in Chap. 5).

6.3.1 *The Most Important Elements of Successful Research Teams*

Although there is no fixed "recipe" for becoming part of a collaborative team, twelve elements of successful research teams have been repeatedly identified (Box 6.1, expanded from the shorter list in Bennett et al. (2010). Trust always comes first on every list. As part of a research team and of much broader communities of researchers, you need to trust yourself and others on the team, and trust the value of, and contribute fully to, group efforts. Remember, you are all working together towards a common goal. The better everyone works together, the more "successful" you and everyone else on the team will be.

Box 6.1 The twelve most important elements of successful research teams

1. Trust
2. Vision
3. Self-awareness
4. Leadership
5. Mentoring
6. Allowing for evolution and dynamics
7. Open, honest, and respectful communication
8. Sharing recognition and success
9. Taking advantage of conflict and disagreement to expand thinking and stimulate research
10. Navigating and leveraging broader networks
11. Equity in support and expectations
12. Learning, teaching, and using practical skills

6.3.2 *Roles on a Team are Constantly Changing*

There are many different ways you can and will participate in a research team. Not everyone on the team will have the same skills or contribute in the same way. At different times, different people on the team will be intellectual leaders, people who "get things done," social coordinators, teachers, and learners. You may start your research experience as a "newbie" doing a lot of learning, but within days or weeks you'll be teaching others and taking on more responsibilities. You can also learn from others who are not STEMM researchers. Support staff, administrators, and friends are also parts of your communities and support systems. If you acknowledge these different roles and recognize that they are not fixed, you will be well on your way to being a successful and valued teammate and collaborator.

6.3.3 *The Importance of Formal and Informal Communication*

If trust is the key to creating successful teams, open and honest lines of communication are the keys to maintaining them. In formal team meetings, you should have as much of an opportunity to share your ideas and opinions as any other team member. Just as you should hear and listen respectfully to your team-members' ideas and opinions, so, too, they should respect yours. But informal communication in unstructured breaks or other social interactions can be even more valuable (Pentland 2010).[3] So, take up or initiate opportunities to join a lab outing, field trip, or other activity outside of the day-to-day research work.

6.3.4 *Teams Can Be Scary and Intimidating*

Joining a collaborative research team for the first time can be difficult and stressful, especially if you have never done research before or are ambivalent for whatever reason about pursuing a STEMM career. You may feel intimidated by others in your research group for many reasons, including:

- You think you aren't as well-prepared as others in the group;
- You're unclear about or misunderstand the team structure or how it "works;"
- You're unsure about "appropriate" personal conduct;
- You are shy or inexperienced in sharing ideas with a group, or you may not be fluent in the primary language of most of the other members of the research team;
- You may have been brought up with and internalized cultural attitudes about speaking up, identifying difficulties ("complaining"), confrontation, or talking about feelings;
- You may not have equitable space or "airtime" because other team members dominate meetings or other avenues of communication;
- There may be too much of a focus on efficiency and work and a lack of informal, lighthearted, or other social elements among team members.

It's important to remember that everyone else on the team had these fears, too, when they started out. Many of us live with them every day, but welcoming, supportive mentors and others in your support systems help all of us move beyond fear and intimidation as we become successful scientists (Vignette 6.1).

Vignette 6.1 Finding support from your network while doing research

Contributed by Selina Cheng

I came into the REU program at the Virginia Coast Reserve LTER as a freshly-declared Environmental Sciences major, lacking any previous field or ecological research experience. I could not have foreseen how much this internship would jump-start my career. What started out as an REU project soon turned into a senior thesis, and then a manuscript. Time and time again throughout this process, my research mentor and lab-mates graciously gave me their undivided attention, taking the time to teach me.

When I think of my undergraduate research network, I think of weekly lab meetings where graduate students gave me constructive feedback on my presentations. I think of one-on-one meetings with my research mentor, learning how to clean and analyze my data. I think of going on runs with my REU friends and talking not just about life but also about science and our research. And when it came time for me to think about life after college, each person in my lab gave me thoughtful insights about their own paths to graduate school and beyond. Now, post-graduation, I still keep in touch with them, grateful to the first group of scientists who surrounded me and nurtured my budding scientific career.

6.4 Becoming Part of Larger Professional Networks

We all need professional network(s) for a variety of reasons, so if you're considering a career in STEMM, you should start building yours sooner rather than later. By professional networks, we mean networks that are larger than your local research team. But like your research team, your professional network is made up of the people who share your enthusiasm for science and research, can help you broaden your thoughts and idea, be a support system in the "work" or "professional" environment, and become scientific collaborators in the future. How do you start to develop your professional network? Show up, engage with others, and get involved.

6.4.1 Building and Expanding Your Own Network

Opportunities to meet people present themselves all the time—in your classes and research groups, clubs organized around particular interests, at office hours held by instructors and professors, and at in-person and online workshops and seminars.

You will find additional opportunities in your undergraduate research experience to meet other students and scientists and build your network. These additional opportunities include meeting visiting researchers who present seminars for you and other undergraduates, and workshops in science skills taught by departmental staff, administrators, and other professionals. Although you should think of almost every aspect of your research experience as an opportunity to meet and engage with new people who *could* become part of your professional network, don't expect that every interaction will lead to something fruitful. Rather than always asking how someone you meet can help you along in your career, you should focus on learning the process of how to engage with other scientists in ways that are meaningful to you and help you understand yourself better.

6.4.2 The Values and Pitfalls of Social Networks

Social networks and other online spaces are other places to meet people and develop your professional network.[4] For every STEMM field, research discipline, or area of emerging interest, you can find an online community of like-minded individuals. In our own experience, social networks can be very interesting and engaging but also time-consuming and insular. Social networks introduce us to many new ideas but also reinforce our existing biases and prejudices.

Finally, you should not assume that what you post on your online social network platforms is private. Potential research mentors, undergraduate research experience program directors, and other employers can use their own social network accounts to learn more about you and your social networks. It's worth considering whether you want what you post to be visible to potential employers or whether you should have multiple accounts for different parts of your virtual life.

6.5 Take-Home Messages

- ✔ STEMM communities are diverse, inclusive support systems.
- ✔ Trust is the key to building successful teams and research communities.
- ✔ Start building your professional network now.

References

Bennett, L. M., Gadlin, H., & Marchand, C. (2010). *Collaboration and team science: A field guide*. Bethesda: National Institutes of Health.

Guimerà, R., Uzzi, B., Spiro, J., & Nunes Amaral, L. A. (2005). Team assembly mechanisms determine collaboration network structure and team performance. *Science, 308*, 697–702.

Institute of Medicine. (2007). *Rising above the gathering storm: Energizing and employing America for a brighter economic future*. Washington, DC: The National Academies Press.
Pentland, A. (2010). The new science of building great teams. *Harvard Business Review, 90,* 60–70.
Wuchty, S., Jones, B. F., & Uzzi, B. (2007). The increasing dominance of teams in production of knowledge. *Science, 316,* 2036–1039.

Part III
After Your Research Experience

During your research experience, you began to form, and have become embedded in, networks of STEMM professionals. These networks can open doors to new opportunities for you. The three chapters in this final part of the book focus on how to take advantage of new short- and long-term opportunities that doing research has opened up for you. We also provide guidance on how to share your research and experience to a variety of audiences through informal channels, in more formal publications, and at professional meetings and conferences. We conclude the book by encouraging you to "pay it forward" by becoming a mentor to the next cohort of undergraduate researchers.

Chapter 7
Continuing the Research

Abstract New questions and directions that were identified during an initial research experience can be used to prepare for another one. This chapter provides guidance on how to share one's research and experience to a variety of audiences through informal channels, in more formal publications, and at professional meetings and conferences. This chapter also discusses how to use a professional network effectively to identify and participate in additional opportunities in STEMM while still an undergraduate: in the next semester, winter or summer term, or year.

While you have been doing your research (Chap. 4) and learning new technical and professional skills (Chap. 5), you have also become part of communities and networks of STEMM professionals (Chap. 6). This chapter focuses on how you can use these networks effectively to identify and participate in additional opportunities in STEMM in the short-term and while you are still an undergraduate: in the next semester, winter or summer term, or year. The next chapter looks further ahead: to graduate or professional schools, jobs, and careers in STEMM.

Your initial research experience can be used to prepare you for another one. New questions and directions you identified during your research also can lead you to create a new research opportunity for yourself. We provide guidance on how to share your research and experience to a variety of audiences through informal channels, in more formal publications, and at professional meetings and conferences. There is one Text Box and one Vignette in this chapter. Box 7.1 encourages you to take some time to reflect on, and evaluate the personal impact of, the research experience you just completed before you plunge headlong into your future. Vignette 7.1 illustrates the power of conversation for sharing your research with others.

7.1 How Did It Go?

It's time to take stock of your research experience. Inevitably, there were parts of it that you loved, others that you hated, and still others that stirred little or no emotions. One of the most important reasons to do a research experience is to understand and

A. M. Ellison and M. V. Patel, *Success in Navigating Your Student Research Experience*, SpringerBriefs in Education, https://doi.org/10.1007/978-3-031-06641-2_7

learn how research is done. Now that you've got some of that understanding, can you see doing more STEMM research, with both its ups and its downs?

You don't have to decide right away, but consider what you got out of your research experience. Maybe you were able to learn some skills and begin to understand the process of research, but did not have the chance to really "do" research. Or maybe you really enjoyed doing research but you'd rather do research outside of STEMM. Or… It's OK to keep exploring.

7.1.1 *Take Time to Reflect*

When you were thinking about doing a research experience, you took time to consider whether and why you wanted to do it, and what kind of research experience would be the right one for you (Box 1.1 in Chap. 1). Before you started your research experience, you answered some questions about your own skills and how you would measure success in it (Box 3.2 in Chap. 3). We suggest that you should close these loops by revisiting these questions. But don't look back at your earlier answers yet. To start, answer the questions in Box 7.1 and then go back to your answers from Chap. 3. What do the similarities and differences in your answers tell you about your accomplishments in research, education, and working in STEMM communities? Once you're ready to think about longer-term steps in STEMM (Chap. 8), we'll return to the questions in Box 1.1.

Box 7.1 Some questions to ask yourself after you finish your undergraduate research experience

1. Was I effectively mentored in doing research?
2. Did I enjoy working on a research team?
3. Did I feel like a respected member of a scientific research team?
4. How well prepared am I to collect, analyze, write up, or present scientific data and results?
5. What are my short-term (1–3 years) goals for employment or additional education?
6. How likely is it that my short-term goals involve research in a STEMM field?
7. How confident am I in my answers to the previous questions?
8. Am I looking forward to another research experience?

Now, compare these answers to how you answered these questions, and any others you may have added, in Box 3.2 *in* Chap. 3. *What changes do you see?*

7.2 What's Next?

If you've gotten this far, you're ready to learn more and do more research in a STEMM field. First, you should finish up your research experience. You're probably asking yourself "the research experience is over and I spent my last stipend check, so what's left to do?" The short answer is, there's plenty left to do. A research project isn't really complete until you've presented your results to your STEMM peers and colleagues and archived your data so that others can use them in the future.

7.2.1 Tying up Loose Ends

Chances are you wrapped up some aspects of your project, but not others. You might have written an abstract of your results and prepared a poster (Faulkes 2021), but have you presented the poster to anyone other than the people in your research group or undergraduate research program? Do you want to? There may be opportunities to present the poster at a group program for students in STEMM at your home institution, or to colleagues at regional or (inter)national meetings. Perhaps the results of your project can—and will—be included in a more expansive paper you work on with your mentor or others in your research group. Alternatively, if you do some additional analysis or collect more data (and analyze them, too), you might be able to publish your results as a stand-alone paper.

7.2.2 Data Management and Data Archiving

If your data aren't clean and organized, you're not going to be able to analyze, interpret, and present them. So before you even think about moving forward with presenting your research, deal with the data. Although researchers in different STEMM disciplines may collect data using different methods, all datasets have a similar underlying structure: observations about and measurements of individual entities or objects that you studied that can be organized in "spreadsheets" (Table 7.1).

Simply putting your data into a spreadsheet, though, is not enough. You also need an accompanying document (which is called "metadata") that describes your dataset in more detail. Every STEMM discipline and subdiscipline has its own standards and expectations for metadata, but all expect at least information about who collected the data, when and where they were collected, and how they were collected (the detailed methods). Variables need to be described in detail. For example, in Table 7.1, what kind of "frog" and "salamander"? How was length and width measured: top-of-the-head to tip-of-the-tail, or some other way (frogs have no tails but salamanders do)? What were the units of measurement: centimeters, millimeters, or something else? How accurate and precise were your measurements? How did you determine color?

Table 7.1 A basic spreadsheet. In this example, the objects studied were different species of amphibians and the measurements and observations included the length, width, and color of each species. Each species gets its own row and each type of observation or measurement gets its own column

Species	Length	Width	Color
Frog	5	3	Red
Salamander	10	1	Blue
...			

Take the time to organize and document your data so that you can easily use them later and so that others can use them in the future.

Finally, it is essential that you digitally archive your well-organized data. We emphasize that data should be archived in a public archive that is "permanent" (has a reasonable likelihood of outlasting your or your mentor's career), that will assign a persist digital object identifier (DOI) to the data, and that is indexed and findable by search engines.[1]

7.2.3 *Sharing Your Research with Peers and Colleagues*

For any practicing scientist, communicating the excitement we feel when we're doing science and getting the word out about our latest findings and their broader importance is as relevant as writing proposals, designing experiments, and collecting and analyzing data. In short, your project isn't really finished until you've told other scientists about your results. Where we might have once hewed to the dictum of "publish or perish," now we must "communicate or evaporate." Contemporary science communication ranges from instant and informal messaging laced with emojis and memes posted on Twitter, Instagram, Facebook, or TikTok, to more considered and formal communication of op-eds, posters, and oral presentations to our colleagues at workshops and meetings, edited or peer-reviewed magazine and journal articles, and full-length monographs and books.

It can be both exciting and daunting to have so many options to communicate the results of our research. And it is crucial for all scientists to learn and become adept in using several different modes of communication. You are not likely to use all these informal or formal modes of communication, but all successful scientists use some of them. You should, too.

Formal communication

When it comes to communicating *within* the scientific community, our communication is most often "formal." By formal, we mean that it conforms to standard methods, styles, and ways of sending and receiving information that are expected and understood by your professional community.

The most important method of communicating with our STEMM peers and colleagues is the peer-reviewed article published in a technical journal (Gastel and Day 2016; Heard 2016). Although every STEMM field has its own particular writing and publishing style, all of them share a commitment to peer review: the critical, often anonymous evaluation by specialists in the field of the methods, results, and significance of the written work. The value of constructive criticism and peer review in improving presentation and clarity of scientific papers and books is rarely disputed, but there also is ample evidence that there are many biases in the peer-review process (e.g., Haffar et al. 2019; Harris et al. 2017). The scientific community continues to explore and develop new methods of peer review to reduce these biases (e.g., Brodie et al. 2021; Tomkins et al. 2017; Tóth 2020; Wolfram et al. 2020).

It's rare that a research project done during a single, short research experience results in a peer-reviewed journal article. It is more common that you can go to a meeting or conference and present your results and ideas for future research that could build on your research experience (Vignette 7.1). Presenting your work at meetings and conferences not only is a great way to communicate the results of your research and your excitement about them, but also is an opportunity to make new connections, build your professional network, and find new opportunities to continue your path in STEMM.

All these formal ways of communicating your results take time—sometimes months or years—and it might feel like they're not worth the effort. Banish that feeling. Formal modes of communicating results of scientific research are crucially important for three reasons.

1. Abstracts, papers, and books—whether actually printed on paper or existing only online—are *the* record or scientific knowledge from which future research will be derived.[2]
2. Formal scientific communication is more long-lasting than are modes of informal communication.
3. Your past, present, and future mentors and employers are more likely to understand and be able to help you learn the mechanics and processes of scientific writing and formal communication than to give you advice on effectively using Twitter, TikTok, or other social-media channels.[3]

It's really important that you see your work through to some kind of publication, whether formal or informal. If you don't communicate the results of your research to your peers and colleagues, you might as well not have done it at all.

Vignette 7.1 Sharing your research

Contributed by Kyla Knauf

I never would have guessed that my prairie research would take me to Hawaii! Maybe I would've pursued science even earlier if I knew!

The summer after my sophomore year, I secured funding to present my REU [US NSF Research Experience for Undergraduates] research at the SACNAS [Society for Advancement of Chicanos/Hispanics and Native Americans in Science] Annual Meeting in Honolulu, Hawaii. Although I was tempted to make this experience a beach vacation, instead I spent a lot of my time pacing a hotel room, repeatedly rephrasing and refining my pitch for my presentation for the next day. When the time came to present, surprisingly, my nervousness was replaced with awe, as I was absolutely stunned by all the positive feedback and meaningful conversations I had with peers and professionals alike (talk about imposter syndrome!). Sharing my research with others at SACNAS truly made me feel part of the science community, as everyone I met treated me with kindness and as an equal. The feedback allowed me to refine and improve both my poster and communication skills. It also built a foundation of newfound confidence that helped me throughout the rest of my undergraduate career and motivated me to pursue graduate school. Scientists in the "real world" almost always work together to make new discoveries and solidify new ideas, so don't be afraid to share your research and start collaborating!

Informal communication

Formally communicating your research is not the only way to share both your science and your experience as a young scientist. Communicating scientific results to broad audiences of interested non-specialists is no longer just done by science journalists. There are many more ways to do "sci-comm" (e.g., Bucchi and Trench 2021). You can write blogs, produce vlogs, and use social media to talk about your research experience and to communicate your enthusiasm about science and the results of your research to colleagues in STEMM *and* to non-specialists. Unlike writing an article or giving a talk or poster, postings on social media are not peer-reviewed or otherwise vetted for accuracy. But it is a lot faster and, increasingly, is a much more effective and efficient way to let other people know what you've been up to.

Informal communication channels also can lead to unexpected opportunities. Off-the-cuff conversations and interchanges started by responses to short postings may lead you to pursue new directions or collaborations in your research, unexpected funding opportunities, or new career paths.

7.3 Doing Another Research Experience

If you do want to do another undergraduate research experience, take stock of what you gained from the one you just finished and what you are left with still wanting or questioning (Box 7.1). Continue this stock-taking by discussing your goals and options with your mentor(s). They can help you consider various different options to continue with research.

In the short term, you might be able to continue working with your research team. Your work may lend itself to a year-long independent research project or undergraduate thesis. Such longer-term projects could by done with your current research mentor or with a new one, at your home institution or another one, or supported with an independent undergraduate research fellowship.[4] But just like doing research itself, finding and securing new research opportunities is an iterative process. Don't just continue your research or do another research experience for the sake of doing it. Rather, find an opportunity that will challenge you and let you grow in new ways.

7.4 Take-Home Messages

✔ Finish your undergraduate research experience by communicating your results to your peers and colleagues.
✔ Take time to reflect on your undergraduate research experience.
✔ Use your network and your communication skills to find your next research opportunity.

References

Brodie, S., Frainer, A., Pennino, M. G., Jiang, S., Kaikkonen, L., Lopez, J., et al. (2021). Equity in science: Advocating for a triple-blind review system. *Trends in Ecology & Evolution, 36*(11), 957–959.

Bucchi, M., & Trench, B. (Eds.). (2021). *Routledge handbook of public communication of science and technology* (3rd ed.). London, Routledge.

Faulkes, Z. (2021). *Better posters: Plan.* Design and present an academic poster. Exeter: Pelagic Publishing.

Gastel, B., & Day, R. A. (2016). *How to write and publish a scientific paper* (8th ed.). Santa Barbara: Greenwood.

Haffar, S., Bazerbachi, F., & Murad, M. H. (2019). Peer review bias: A critical review. *Mayo Clinic Proceedings, 94*(4), 670–676.

Harris, M., Marti, J., Bhatti, Y., Watt, H., Jand, Macinko A., & Darzi,. (2017). Explicit bias towards high-income country research: A randomized, blinded, crossover experiment in English clinicians. *Health Affairs,36*(11), 1997–2004.

Heard, S. B. (2016). *The scientist's guide to writing: How to write more easily and effectively throughout your scientific career*. Princeton: Princeton University Press.
Tomkins, A., Zhang, M., & Heavlin, W. D. (2017). Reviewer bias in single- versus double-blind peer review. *Proceedings of the National Academy of Sciences, 114*(48), 12708–12713.
Tóth, J. (2020). Blind myself: Simple steps for editors and software providers to take against affiliation bias. *Science and Engineering Ethics, 26,* 1875–1877.
Wolfram, D., Wang, P., Hembree, A., & Park, H. (2020). Open peer review: Promoting transparency in open science. *Scientometrics, 125*(2), 1033–1051.

Chapter 8
Building on the Experience

Abstract This chapter focuses on continuing to do STEMM research beyond the undergraduate degree. Because most STEMM professionals do not have a post-graduate degree, this chapter presents and discusses a range of career paths that provide opportunities for research. Deciding which path to pursue is suggested by a set of self-reflective questions to keep in mind when considering longer-term career paths in STEMM.

Ask yourself what you're going to do after you finish your undergraduate degree. Is your answer that you're going to "take time off" and get a job before... well, before what? For many undergraduates in STEMM fields, especially those who have done a college- or university-based research experience, you might think that the next step after completing a BA or B.Sc. is to go to graduate or professional school. Thus, the parallel to taking "time off" to work after finishing college is to "go back to" school for another degree. The implication of taking "time off" and then "going back to" school implies that the one true path is an advanced degree and the subsequent career that requires it. This idea may have been emphasized to you by the discussion of "alternative careers" by professors and other academic mentors, for whom any job other than a professorship is a lesser "alternative."

Nothing could be further from the truth (Fig. 8.1).[1]

This chapter focuses on continuing to do STEMM research after you complete your undergraduate degree. We start by illustrating a range of possible career paths other than the academic path that was most likely suggested to you by your university-based undergraduate research mentor: a B.A. or B.Sc. → M.Sc. and Ph.D. → university post-doc → tenure-track position, and finally, to becoming a tenured academic faculty member just like your mentor. The Text Box (Box 8.1) includes a list of questions to keep in mind as you think about longer-term career paths in STEMM.

We discuss how to use your undergraduate research experience(s) to your best advantage when you are looking for post-graduate internships or full-time employment in STEMM fields. We also give a general overview of how to think about, apply to, and prepare for graduate studies in masters (M.A. or M.Sc.) or doctoral (Ph.D.) "academic" programs and professional schools. Vignette 8.1 shows how there can be unexpected and exciting outcomes from undergraduate research.

A. M. Ellison and M. V. Patel, *Success in Navigating Your Student Research Experience*, SpringerBriefs in Education, https://doi.org/10.1007/978-3-031-06641-2_8

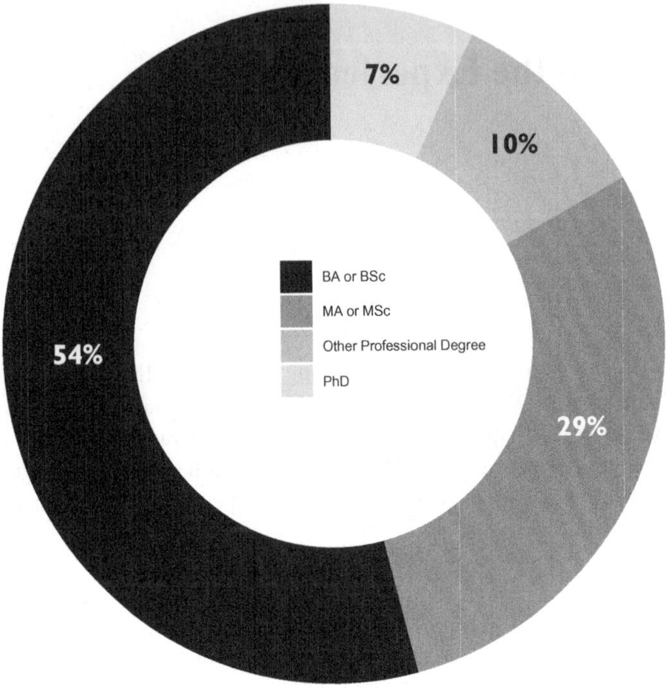

Fig. 8.1 Of the more than 15 million graduates of two- and four-year colleges in the USA who have a BA or B.Sc. in STEMM, only about a quarter of them also have a have a MA or M.Sc., and less than a tenth have a PhD (Milan 2019)

8.1 A Range of Career Paths

While you were doing your research experience, you probably met other people at many different educational or career stages in your STEMM field. Did these people represent a narrow or wide range of career possibilities and role models for you? And why would it matter? How can your experience and the people who are now part of your professional network help you think about your long-term goals?

8.1.1 Long-Term Thinking

In Sect. 7.1 in Chap. 7, we suggested that right after you completed a research experience that you should take time to reflect on what you learned and accomplished (Box 7.1). But now, as you're ready to move on from undergraduate research to a career in STEMM, your self-reflection should focus on the longer term.

The first thing you did when thinking about undergraduate research was to consider why you wanted a research experience and how it would fit into your short-term goals (Box 1.1). In Box 8.1, we update those questions to help you think about your longer-term goals for work in STEMM and how continuing to do research can help you meet those goals.

Box 8.1 Self-reflection: Where might I go next in my STEMM career?

For many students, a research experience solidifies their goal for a career in STEMM. For others, it leads to a decision to look in new directions. Which way do you want to go for the next few years? Either one is a positive outcome of your research experience!

First, ask yourself:

- Is now the time to commit to a career path? The answer is not a simple yes or no. Rather, you should ask: What are my long-term (5–10 years) goals for employment or additional education?

Then, ask yourself:

- What do I need to do next to get started on the path that will lead me to my goals?
- What are the multiple ways I learn effectively?
- Am I ready to keep pushing myself outside of my comfort zone?
- Would doing more research help me take the next step on my career path?

If the answers to these questions point you toward further research or a career in STEMM, then this next set of questions will help you choose among different ways to do that.

- Do I need a paying job on a regular basis?
- Am I okay going far away or do I want or need to stay close to home or school?
- What type of support do I want from the "research community?" For example, do I want to work one-on-one with a more senior researcher (mentor) or do I want to work in teams with other students?
- Do I learn better working informally at my own pace or in a structured internship or program?
- Would I thrive in a more or less diverse environment?

Now, compare these answers to how you answered these questions in Box 1.1 in Chap. 1 of this book.

- Should I stay the course or reorient my path?

8.1.2 Think Outside the Box

If you were doing research at a college or university, chances are that almost everyone you worked with or ran into was another undergraduate, a research assistant, a graduate student, a post-doc, or a professor. If you were doing research at a scientific company, chances are that your daily professional network was made up of other interns, scientists, and other employees of that company or other companies doing similar work. Similar narrowness would occur if your research experience was based at a government lab or a non-governmental organization (NGO).

Seeing outside any of these initial boxes matters because you are just getting started in STEMM research. Finding new opportunities and getting a job in any STEMM sector takes time and work, and in any of them, you'll be competing with others for a limited number of positions. You also need to realize and appreciate that the expectations and measures of productivity or success in each sector can be very different. Publishing research papers with open, archived data and code may be much more important if you're interested in an academic position than if you're interested in doing research with a pharmaceutical company that is focused on product development, patents, and profits. Other potential employers may value "white papers" or technical reports more than citation counts or journal "impact factors."[2] Taking the time and making the effort to learn the similarities and differences in expectations of different STEMM career sectors will expand your horizons, increase the number of opportunities you'll have as a STEMM professional, and in a competitive arena, set you apart from the crowd.

The most valuable and important aspect of getting research experience in STEMM is that you learn how to think critically. Knowing how to ask questions and seek their answers is a fundamental skill that can be used in any job or career—in STEMM and in other fields, too. Thus, the range of opportunities available to you in STEMM is limited only by your imagination, enthusiasm, and drive, not by the specific field or subfield in which you had your research experience or in which you are getting (or already have) your undergraduate degree.

8.2 Employment: It's Not "Time Off"

You will be much more successful if you view a job in STEMM as the next step on your continuing career path. One or more research experiences can give you a competitive edge in getting a job doing research in STEMM. Your potential employers will expect you to take the job seriously, not to use it simply as a placeholder while you figure out what you "really" want to do with your life. If you are more interested in short-term or temporary positions to gain additional experience or help provide you with additional career focus, consider doing an internship. But whether you are looking for internships or permanent jobs, don't assume that either is a bridge "back to" school. As a STEMM researcher, you are always learning, whether you're in school or at work.

8.2.1 *Internships*

An internship is a fixed-term job that many businesses, non-profit organizations, government agencies, and NGOs offer to individuals who are interested in expanding their skill set or gaining additional experience. They differ from the research experiences we have been discussing throughout this book in that they rarely have educational components and they put less emphasis on community-building activities. STEMM internships usually are open both to current students and recent graduates. They also pay salaries (as opposed to "stipends")[3] but are unlikely to cover costs of housing, food, or travel.

Doing one or more internships is a great way for you to develop more technical skills and to give you more exposure to the day-to-day nitty-gritty of STEMM research. In some cases, internships can lead to permanent positions at the same organization or a related one.

8.2.2 *Scholarships and Fellowships*

There are many scholarship and fellowships available to students who are about to, or have recently received, their undergraduate degree. For non-USA residents and citizens interested in studying or doing research in the USA, the Foreign Fulbright Program offers one of the most extensive set of opportunities for students in all fields, including STEMM. For students from any country, including the USA, we recommend searching and browsing the ever-increasing number of websites devoted to listing similar opportunities.[4]

8.2.3 *Jobs*

As we noted at the beginning of this chapter, more than half of the graduates from 2- or 4-year colleges never attend graduate or professional schools (Fig. 8.1; Milan 2019). Rather, the most common path for a recent college graduate is that they look for a job in their area of interest. Indeed, more than 80% of graduates in STEMM fields get jobs (or create their own jobs or companies) in STEMM, or that take full advantage of the critical-thinking skills that they developed through their studies (Milan 2019).

There are plenty of people who can advise you and help you find your first job after college, and there are plenty of guidebooks offering advice. Although this is not one of those books, we do have few pointers to offer, and they really don't differ from what we presented in Chap. 2 of this book:

- Reflect on your goals;
- Match opportunities to your goals;
- Put care into your applications;
- Network;
- Be persistent.

Emphasize not only that you did one or more research experiences but also, and more importantly, what you learned or produced from them that could benefit your potential employer.

Most importantly, in whatever work you do, you will carry with you the ideas and skills you learned from doing research in STEMM.

8.3 Post-Graduate Research and Education

You may decide to continue your education in STEMM beyond your undergraduate degree. Such post-graduate education ranges from post-baccalaureate ("post-bacc") programs leading to certificates documenting your mastery of specific skills or completion of STEMM courses; "graduate" degree (typically a MSc or a PhD); or a professional degree (from the alphabet soup of MD, MFS, MES, MPH, the various permutations on engineering degrees, etc.). What's the difference and how do you decide which ones are right for you?

8.3.1 Post-Baccalaureate Programs

A number of colleges and universities offer one- or two-year post-baccalaureate ("post-bacc") certificate programs in STEMM fields. It is important to note that, as a rule, post-bacc programs are course or skills-based and lead to a certificate or assertion of qualifications, not an additional degree.

In the USA, post-bacc programs can be found in all fields, including all STEMM fields, but most of these programs are focused on biomedical careers.[5] In contrast, in Europe and the UK, post-graduate "conversion" and vocational courses, certificates, and diplomas are offered by many schools to students who are looking to learn skills or take classes in a field other than that in which they received their undergraduate degree.[6]

8.3.2 *Different Types of Graduate Degrees*

Typically, a PhD ("Doctor of Philosophy") is an "academic" degree that requires the completion of a substantial, original research project. Most PhD students are interested in pursuing a career in research, usually in an academic institution, but also perhaps in a government laboratory, or an private-sector organization that has a research arm. The metrics of success for these types of positions are the quality and number of peer-reviewed papers published plus the number and amount of grants competed for and awarded to support the research. For positions that require a substantial amount of teaching, PhD students should also be sure to get teaching experience.[7]

In contrast, professional degrees, awarded by professional schools, such as medical school, schools of public health, engineering schools, etc. emphasize learning complex skills and putting them into practice. Except for programs that lead to simultaneous awarding of a professional degree plus a PhD (e.g., a MD/PhD), research is not usually emphasized or encouraged in professional degree programs. Most students with professional degrees are interested in careers in a particular profession (medicine, engineering, public health, environmental science, *etc.*) where the metric of success is how effectively and successfully they can put their skill-set into practice.

Finally, an "academic" masters degree (MA or MSc) can work in either way. For many professions, a MA or MSc is either required, strongly recommended, or is your ticket to advancement and a better salary. Some PhD programs require an MA or MSc for admission to the doctoral program, whereas others award a masters degree following completion of the qualifying examinations (typically taken in the second year of a PhD program).

8.3.3 *Graduate or Professional School: Which One's for You?*

There are a lot of things you should think about when considering whether graduate or professional school is right for you. Up first is what field you're interested in. Different STEMM fields differ in their focus on "basic" and "applied" research.[8]

Second, you should ask yourself whether it is the right time of your life to attend graduate or professional school. All of us have different immediate needs or obligations that may complement or compete with our medium- or longer-term goals (Box 8.1). Take care of what you have to do first and don't worry about missing the graduate- or professional-school boat. As you learned in your undergraduate research experience, research always opens up more questions than it answers, so the time and opportunity to do research in graduate school will always be there.

Once you've decided on a field of interest and that the time is right to pursue a graduate or professional degree, then you should take the time to consider your goals, needs, and desires for graduate studies. For example, you can extend the questions you used to consider whether to pursue further undergraduate research opportunities (Box 7.1 in Chap. 7) to post-graduate research.

Those questions are still applicable because in many ways, your undergraduate research experience was a good introduction to what you will be doing in graduate school in STEMM. You need to do the same mental preparation and logistical work to find, apply to, and get ready for graduate or professional programs (Carson et al. 2021) as you did for your undergraduate research experience (Chaps. 1–3).[9] Similarly, all the skills you developed during your undergraduate research experience (Chaps. 4–6) will serve you well and be extended while you are working to successfully complete a graduate degree.

Finally, keep up with the professional networks you built and developed as an undergraduate researcher and that nurtured you along the way (Vignette 8.1). Share your research with them while drawing on them for different kinds of support. Some of the people in your network could give you new perspectives and insights about graduate school and beyond. Others might be in a position to write you strong letters of reference to accompany your applications. Still others could be potential graduate advisors and mentors.

Vignette 8.1 Unexpected long-term outcomes from your undergraduate research experience

Contributed by Karen T. Y. Tang

One month into my undergrad, I received an email out of the blue inviting me to work as a research assistant. Little did I know that that one single email would change the rest of my life. That research opportunity led to working for another research lab, which then snow-balled into a three-month-long paid internship in Germany, and finally to an Honor's thesis in addiction psychology, where I found my true passion in research. I have cherished all these memories, from conducting research in Europe and finding life-long collaborators and friends, to publishing five peer-reviewed articles as an undergrad, including a "Best Paper of the Year" awarded by the International Society for Justice Research for an article that was rejected three times. I credit that single auspicious email in getting me into the highly competitive field of clinical psychology, where I am currently training to be a scientist-practitioner. As someone who is BIPOC and disabled, I never would have discovered my joy for research had I not taken a leap of faith in my first year as an undergrad.

Go ahead and take a chance on a research opportunity. It's waiting right there for you. You simply have to reach out and seize it.

I dare you.

8.4 Take-Home Messages

- ✔ Start now to define your longer-term goals for a STEMM career.
- ✔ Recognize that working in STEMM is part of your career path.
- ✔ Find what will be best for you if you decide that you want to pursue an advanced degree in STEMM.

References

Carson, W. P., Kuebbing, S. E., Betras, T. L., Campbell, A. S., McQueen, E. W., Moore, C. L., et al. (2021). Advice on applying to graduate school in ecology and evolutionary biology: How to prepare and a step-by-step guide. *The Bulletin of the Ecological Society of America, 102*(4), e01917.

Milan, L. (2019). Characteristics of college graduates, with a focus on veterans. Retrieved December 02, 2021, from https://www.nsf.gov/statistics/2019/nsf19300/

Chapter 9
Becoming a Mentor

Abstract The final chapter in this book focuses on "paying it forward" to the next generation of STEMM professionals. Following a successful undergraduate research experience, students should find their own voices, embrace their knowledge and experiences, overcome the continual challenge of "impostor syndrome," and recognize the value of sharing their experience with others as a peer mentor.

Through doing mentored undergraduate research, you have gained a deeper understanding of research and education in STEMM, learned a variety of new skills, and become part of networks of like-minded professionals. To close out this book, we ask you to take time to reflect on the relationship you had with your mentor and consider making the transition yourself from being a mentee to becoming a mentor to other young scientists. We encourage you to find your own voice as a STEMM professional, embrace your knowledge and experience, overcome the continual challenge of "impostor syndrome" (Vignette 9.1), and recognize the value of sharing your experiences with others.

9.1 The Importance of Mentorship

STEMM theorists and practitioners have argued that the association between a mentor and a mentee is the most significant collaborative relationship in any scientist's career (Ma et al. 2020). The US National Academy of Sciences has called mentorship a series of "high-stakes, interpersonal encounters and exchanges" (National Academies of Sciences, Engineering, and Medicine 2019, p. x) that acts as a "catalytic factor to unleash individuals' potential for discovery, curiosity, and participation in STEMM" (National Academies of Sciences, Engineering, and Medicine 2019, p. x). But not every intellectual partnership becomes a mentor-mentee relationship. Some of the people you worked with while you're learning about and doing research became your mentors, but others may have related to you as advisors, supervisors, or managers.[1] The mentor-mentee relationship is a unique one, and becoming a mentor takes time, trust, mutual respect.

A. M. Ellison and M. V. Patel, *Success in Navigating Your Student Research Experience*, SpringerBriefs in Education, https://doi.org/10.1007/978-3-031-06641-2_9

9.2 Reflect on Your Mentee-Mentor Relationship

Throughout your mentored research experience(s), there were almost certainly things that your mentor did that helped you have a positive and productive experience. And as in any relationship, your mentor probably said or did things that were not helpful or were downright irritating. What else could you have expected? You and your mentor are different people, are at different career stages, and have had different life experiences. Although no one gets everything right, we hope that the good and positive outweighed the bad and negative. Did it?

When reflecting about your mentored research experience, ask yourself a few questions:

- Who were (and perhaps still are) my research mentors?
- What interests did/do I share with them?
- How did each of them influence my intellectual and personal development, both as a scientist and a person?
- What barriers did they help me overcome?
- What did I teach (give back to) them?

Which of these questions and answers seem to be most important to you now? Are they the same things that you thought were important when you were just considering whether to do a research experience (if you don't remember those things, take the time to revisit your answers to the questions in Box 1.1 in Chap. 1)? What, if anything, has changed in your outlook?

And perhaps most importantly, could you incorporate the central attitudes and behaviors of mentorship—sharing, influencing, teaching, and helping others to overcome barriers—in your own day-to-day activities, your emerging career in STEMM, and your life? If you can, you're on your way to becoming a mentor to others. At the same time, you will continue to have mentors who will continue to influence your life and learning.

9.3 Becoming a Mentor

There are dozens of definitions of mentors and mentorship. The dictionary definition of a mentor is "a person who acts as guide and adviser to another person, especially one who is younger and less experienced" (OED 2020). In STEMM fields, mentors usually are more experienced, trusted people who can offer guidance, support, and counsel to someone who is less experienced and interested in collaboratively learning about an area of shared interest. Now that you've completed one or more research experiences, you are more experienced than someone who hasn't done a research experience. So, you could be a mentor to them. How do you start?

9.3.1 *The Essence of Mentorship*

The first thing to know is what you're signing up for as a mentor.

> Mentors of new researchers are individuals who see potential in their mentees and empower them to become self-aware, independent-minded scientists and scientifically literate members of their communities.[2]

We emphasize that your main goal as a mentor should be to help and support your mentees, not to advance your own agenda or career. Are you ready for that? It's fine if you're not. For a variety of reasons, many STEMM researchers never become mentors. But if you are, read on.

9.3.2 *Know Yourself*

Earlier in this chapter we asked you to answer a few questions about your research mentors. If you're thinking about mentoring others, ask yourself:

- Am I ready to help my mentees' become scientists and peers?
- What barriers can I help my mentees to overcome? Can I draw on my professional and personal networks to help my mentees overcome barriers that I can't help them with?
- Am I open to learning things from my mentees that could make me a better scientist and person?
- Am I ready to take advantage of being a mentor myself to change my perception of my own mentors and of mentorship in general?

For mentors and mentees, reciprocal and reciprocated trust and communication are key components of their relationship. As in any interpersonal relationship, as a mentor, you and your mentees will need to continually nurture, build, and re-create your ways of working together effectively and productively. But first, you need to trust in yourself to make it work.

9.3.3 *Recognize that You Are Not an Impostor*

Trusting yourself as a scientist, as a STEMM researcher, and as a mentor for, and co-learner with, others can be really difficult. You may doubt your abilities and feel that you're not really a part of the STEMM community and can't ever be truly welcomed into it. Although we cannot eliminate your self-doubts in a short chapter, we can assure you that you are not alone in having these self-doubts. We all have them at various times and career stages (see Vignette 9.1).

Vignette 9.1 The authors of this book reflect on their own "impostor syndrome"

Contributed by Manisha V. Patel and Aaron M. Ellison

Manisha: I remember having an "impostor" moment the day Aaron asked me to co-author this book. We had gone out to lunch and he presented the idea. Initially I thought "This is a great idea. He's great at writing books." Then I realized he was proposing that **we** write this book together. My inner voice said "He thinks I have 'something' to contribute, and wants to write this book together. If I say 'no,' he'll know that I'm a fraud, so I'll say 'yes' and hope that he doesn't realize I'm an impostor." So, I said yes and we shook on it. Up until the moment we started writing the book (and even a few times during), I have felt like an impostor. But perhaps more worrisome for me was that Aaron would realize I was an impostor. It didn't happen because I do have something to contribute, and I'm not an impostor (just feel like it sometimes). I'm sure this is not the last time I will have an "impostor" moment, but it won't stop me from moving onwards.

Aaron: As a first-year graduate student, one of the professors in my department asked me to review a manuscript by a famous ecologist. "Who me," I thought? "I'm not qualified to do this. I've only just started graduate school and I couldn't possibly know enough to comment on this—or any—scientist's work." But I swallowed my doubts and agreed to review the paper. I read it and thought it was terrible. I wrote up my review and sent it in. Peer reviews of articles are meant to be anonymous, but I was sure the author would find out who had criticized their paper, find all sorts of flaws in my review, and make sure I couldn't finish graduate school, much less pursue a STEMM career. Now, almost 40 years later and as a senior journal editor myself, my reviews, comments, and decisions whether to accept or reject an author's paper are no longer anonymous. To this day, I still wait for my colleagues to pan my decisions to have accepted obviously bad papers or authors to appeal my decisions to reject their obviously good ones. I do make some bad decisions—which I own up to—but I also now accept other's assessments that my editorial work has, in the main, been helpful to them as individuals and to the field in general. For me, trusting my colleagues has been the way to overcome my own impostor syndrome.

At their most extreme, however, these self-doubts can have serious negative effects, not only on your ability to do research or your career, but also on your physical and mental health and well-being. Across all the professions, including all STEMM fields, many apparently successful individuals suffer from "impostor syndrome." Clance and Imes (1978) coined the term "impostor syndrome" from their pioneering study of 150 highly successful women who had earned their PhDs in eight fields: law, anthropology, nursing, counseling, religious education, social work, occupational therapy, and teaching. They reported that many of these women attributed their success, accomplishments, and standing in the field to luck and contingent events, not to their actual merit or skills. Since its identification and naming, impostor syndrome has been recognized much more widely. It afflicts individuals of all genders, ethnicities, cultural backgrounds, and career stages (e.g., Pederson 2020), although it is more prevalent in women and minoritized students and professionals (e.g., Arnold et al. 2020; McGill et al. 2021; Vajapey et al. 2020).

Organizational structures, especially in colleges and universities, and criteria for success and reward in many STEMM fields, all can contribute to one's sense of being an impostor (Bothello and Roulet 2019). As researchers, we are encouraged early on to identify "our" unique question or problem, to answer or solve it, and to "own" it. While doing that, it's easy to forget that our work is part of a larger, collaborative effort and that our results have value to others (Bothello and Roulet 2019). This sense of isolation, coupled with the constant, nagging fear of being "found out" as a self-absorbed fraud doing useless work can be either paralyzing or lead one to work constantly towards an impossible level of perfection.

Although we may never banish all of our self-doubts, there are many different approaches to ridding yourself of impostor syndrome (e.g., Clance and Imes 1978; Dickerson 2019; McGill et al. 2021). All of them focus on cognitive and behavioral changes, setting realistic expectations, and recognizing and trusting in your own strengths and capabilities.[3]

9.3.4 Share Your Experiences

One of the most important things a mentor does is to share their experiences with their mentees. You don't share your experiences to brag about how good you are. Rather, you should be sharing with your mentees events and insights that you have learned along your path that helped or hurt you as you developed your research and your STEMM career. In such sharing, you also give voice to the many things that were left unsaid but that you wish you had known earlier.

At the same time, as a mentor you should always recognize that what worked for yourself as a mentee may not work for your mentees. Instead, as a mentor you are consciously giving your mentees a set of tools, skills, and attitudes that they can reshape and use in their own ways and on their own paths. Be ready to listen to your mentees' thoughts, ideas, and concerns, and be open to changing your approach in different situations and contexts.

Throughout this book we have presented and discussed mentors, mentees, and mentorship in a particular context: the undergraduate research experience. As you move forward in your career, from student through new STEMM researcher and to becoming an established professional, you will have other mentors and you may mentor others. It is always up you whether to choose to identify a particular person as a mentor. Similarly, you may set out to be a mentor for your students, but they too will choose whether or not to consider you as one of their mentors. In our companion book, *Success in Mentoring Your Student Researchers*, we delve into the undergraduate research experience from the mentor's perspective.

9.4 Take-Home Messages

✔ Find your own voice as a STEMM professional and share it with others.
✔ Pay it forward and use your experience as a mentee to become a successful mentor.
✔ Embrace your knowledge to overcome impostor syndrome. We all feel it but it shouldn't stop us.

References

Arnold, A. C., Wilkins-Yel, K. G., Bekki, J. M., Bernstein, B. L., Natarajan, M., Randall, A. K., Francies, R., & Okwu, C. E. (2020). Examining the effects of STEM climate on the mental health of graduate women from diverse racial/ethnic backgrounds. *American Society of Engineering Education Papers on Engineering Education Repository*. https://peer.asee.org/34617

Bothello, J., & Roulet, T. J. (2019). The imposter syndrome, or the mis-representaiton of self in academic life. *Journal of Management Studies, 56*(4), 854–861.

Clance, P. R., & Imes, S. A. (1978). The impostor phenomenon in high achieving women: Dynamic and therapeutic intervention. *Psychotherapy: Theory, Research and Practice, 15*(3), 241–247.

Dickerson, D. (2019). You are not an impostor. *Nature, 574,* 588.

Ma, Y., Mukherjee, S., & Uzzi, B. (2020). Mentorship and protégé success in stem fields. *Proceedings of the National Academy of Sciences, USA, 117*(25), 14077–14083.

McGill, B. M., Foster, M. J., Pruitt, A. N., Thomas, S. G., Arsenault, E. R., Hanschu, J., et al. (2021). You are welcome here: A practical guide to diversity, equity, and inclusion for undergraduates embarking on an ecological research experience. *Ecology and Evolution, 11,* 3636–3645.

National Academies of Sciences, Engineering, and Medicine. (2019). The science of effective mentorship in STEMM, Washington, DC: The National Academies Press

OED. (2020). "Mentor", n. OED Online https://www.oed.com/view/Entry/116575

Pederson, T. (2020). A 20-year encounter with the imposter syndrome. *Molecular Biology of the Cell, 31,* 2509–2510.

Vajapey, S. P., Weber, K. L., & Samora, J. B. (2020). Confidence gap between men and women in medicine: A systematic review. *Current Orthopaedic Practice, 31*(5), 494–502.

Notes

Notes for Chap. 1

1. A detailed discussion of "the" scientific method is well beyond the scope of this book. Wikipedia's entry on the scientific method is a good starting point to the hundreds of years of philosophical discussions about it that continue to this day.
2. We note that when a student can do an undergraduate research experience can differ within and among institutions and countries. In some instances, opportunities are restricted to students who are currently enrolled in a 2- or 4-year college (i.e., to students who are enrolled in, or have completed their first, second, or third year of study). Others may be open to students who have completed secondary school and been accepted into, but not yet enrolled in college. Still others may be open to students who have graduated from college but are not yet enrolled in post-graduate studies. See Chap. 2 for additional discussion of application details and requirements.
3. Study abroad programs, including experiential ones such as The School for Field Studies or SEA Semester charge tuition and other costs and offer college credits but rarely have strong, independent research components.
4. Current NSF guidelines stipulate that only undergraduates who are US citizens or permanent residents can receive funding from REU Site grants. But many institutions that support REU Sites also have internal, non-governmental funds that can support international students on student visas or who have work authorization through the USA's Deferred Action for Childhood Arrival (DACA) program. If you are uncertain about your eligibility for an undergraduate research program, email the program's director or coordinator, express interest, and ask about your eligibility and alternative sources of funding.
5. Minority-serving institutions (MSIs) include colleges and universities that have an historical origin in serving minoritized individuals or now have a large percentage of students from minoritized groups. These include Historically Black Colleges and Universities (HBCU), Predominantly Black Institutions (PBI), Tribal Colleges and Universities (TCU), Native American Non-Tribal Institutions (NANTI), Hispanic-Serving Institutions (HSI), Alaskan Native- or Native Hawaiian-Serving Institutions (ANNHs) and Asian American- or Native American Pacific Islander-

A. M. Ellison and M. V. Patel, *Success in Navigating Your Student Research Experience*, SpringerBriefs in Education, https://doi.org/10.1007/978-3-031-06641-2

Serving Institutions (AANAPISI) serving colleges and universities. The US Department of Education maintains a listing of all MSIs in the USA.

Notes for Chap. 2

1. A good starting place to learn more about implicit bias, ways to identify it in yourself and others, and how to minimize it is at: https://implicit.harvard.edu/implicit/takeatest.html.
2. To learn more about how mentors and program directors are working to ensure equity in undergraduate research programs, see Chap. 3 in our companion book, *Success in Mentoring Your Student Researchers.*
3. "CV" is the abbreviation for *curriculum vitae*, literally one's "course of life." A CV summarizes an individual's entire career(s) and accomplishments and has no fixed length. In contrast, a résumé emphasizes one's background and skills relevant for a specific job and rarely exceeds two pages.
4. LibreOffice is an open-source set of software tools for word processing, spreadsheets, and presentations that provides virtually all the functionality of commercial packages, but at no cost to you.
5. There is often a fee charged for an official transcript, and if this fee presents a financial hardship, ask for a waiver. If the application form or system does not specify an "official" transcript, an unofficial one is usually fine.
6. A small example. One of the authors of this book (Aaron) majored in East Asian Studies as an undergraduate. The lack of any chemistry courses on his undergraduate transcript was often flagged and discussed as a potential sticking point when he was applying to graduate school and later for faculty positions at liberal-arts colleges.
7. For additional discussion of how these data are collected and used, see Chaps. 2 and 7 in our companion book, *Success in Mentoring Your Student Researchers.*

Notes for Chap. 3

1. These questions are derived from common survey instruments used to evaluate outcomes of undergraduate research experiences. Examples include the Student Assessment of Learning Gains (SALG), the Undergraduate Research Student Self-Assessment (URSSA) and the Center for the Improvement of Mentored Experience in Research (CIMER) Assessment Platform. We have used a combination of these in our work assessing the Harvard Forest Summer Research Program in Ecology.

Notes for Chap. 4

1. For more discussion of this aspect of the mentor-mentee relationship, see Chap. 5 in our companion book, *Success in Mentoring Your Student Researchers.*
2. To learn more about the importance of early-career mentoring to established scientists, see Chap. 1 in our companion book, *Success in Mentoring Your Student Researchers.*
3. For a more detailed discussion of what your mentor might expect, see Sect. 3.1 in Chap. 3 of our companion book, *Success in Mentoring Your Student Researchers.*

Notes for Chap. 5

1. The definition of responsible conduct of research (RCR) that we use here is adapted from an overview of RCR written by the US National Institutes of Health Office of Intramural Research.
2. Good sources for guidelines on RCR include the US National Institutes of Health Office of Intramural Research, the US National Science Foundation Office of the Director, the American Psychological Association, and the European Network of Research Ethics and Research Integrity.
3. In the USA, these topics will also be covered in additional training courses and workshops focused on compliance with "Title IX." Title IX is a federal civil rights law (Public Law No. 92–318, 86 Stat. 235; codified as 20 U.S.C. §§1681–1688) passed in 1972 that prohibits sex-based discrimination in any school or other education program that receives federal money.
4. The FAIR guidelines—making sure your data are **F**indable, **A**ccessible, **I**nteroperable, and **R**eusable— encapsulate current best practices for responsible and ethical data management.
5. The CRediT system provides clear guidelines for maintaining integrity and transparency in determining and asserting (co-)authorship of research papers. Open-access refers to Creative Commons licensing.
6. In the USA, the Animal Welfare Act (1966, amended through 2008) is administered locally by Institutional Animal Care and Use Committees (IACUC). In Europe, national regulations for animal experimentation apply across the EU but are implemented on a country-by-country basis under Directive 2010/63/EU.
7. In-person training courses for research involving human subjects take at least one or two full days. Online versions take 10–20h to complete.

Notes for Chap. 6

1. Wutchy et al.'s detailed analysis of nearly 20 million research papers published and more than 2 million patents filed between 1960 and 2000 found that research teams produce more knowledge and more highly cited research than do single individuals, and that the proportion of research done by teams increased exponentially over that same period. Their findings applied not only to all STEMM fields, but also to research in the arts, humanities, and social sciences.
2. Guimerà et al. used data from teams of various sizes that had achieved varying degrees of success in the arts and sciences. They identified a process of self-assembly of teams that occurred across fields as a function of team size, the proportion of new individuals, and how likely members of the team were likely to repeat previous collaborations. Like Wuchty et al.'s findings that we discussed in the previous note, Guimerà et al. found a steady increase in team size and performance over time, with more rapid increases in the sciences than in the arts or social sciences. They found that successful teams have a higher fraction of repeat collaborators ("incumbents") who bring continuity of process and knowledge, and that more diverse teams outperform less diverse ones. Their results also suggested that there could be an optimal size for teams, but that this size is likely to vary among disciplines.

3. Pentland presented results from a study in which he and his colleagues deployed wearable electronic sensors to track interpersonal interactions among co-workers in the private sector. They found that productivity increased up to 20% when coffee breaks for all staff were scheduled simultaneously.
4. The landscape of social networks and the available platforms changes so rapidly that we cannot give advice on any specific platforms or social network.

Notes for Chap. 7

1. Permanent and public data archives that many scientists use include Dryad, Figshare, GenBank, EDI, and the wide range of databases used for crystal structure and data. There are many others, but your own website, that of your research group, major department, or school, or GitHub are not considered permanent repostories.
2. You may be familiar with the famous quote often attributed to Isaac Newton (1675), that "If I have seen further it is by standing on the shoulders of giants." Newton developed, among other things, the theory of classical mechanics, which is the foundation of modern physics; and invented calculus with the mathematician Gottfried Wilhelm Leibniz. Indeed, this idea and versions of the quote may predate Newton by over a millennium (see the Wikipedia entry on the quote at: https://en.wikipedia.org/wiki/Standing_on_the_shoulders_of_giants). As you learned in your training course on responsible conduct of research (see Chap. 5), it's crucial that you acknowledge your intellectual debt to other researchers through appropriate citations and other forms of credit.
3. Indeed, you may have had workshops on writing abstracts and papers as part of your research experience (see Box 5.1 in Chap. 5).
4. Many colleges and universities offer their own fellowships and other support for undergraduates who develop independent research projects. There also are a number of more competitive national or international fellowships, such as the Barry Goldwater Scholarship in the USA. The scholarship, fellowship, and financial aid offices at most colleges and universities maintain up-to-date lists of opportunities across the constantly shifting landscape of undergraduate research awards

Notes for Chap. 8

1. The data in Fig. 8.1 are from the 2017 National Survey of College Graduates conducted by the US National Science Foundation. Milan (2019) presents a more fine-grained analysis of these data.
2. In Sect. 7.2 in Chap. 7, we discussed a range of different ways to communicate results of scientific research. As we pointed out there, publishing a peer-reviewed paper in a technical journal is the most important and recognized way to communicate results of research to one's academic colleagues in STEMM. The relative importance of a journal article may be indicated by the "impact factor" of the journal in which it is published or by a variety of "alt-metrics."
   But in government agencies, think tanks, and many non-governmental organizations, "white papers" and technical reports are more highly valued. A white paper is a relatively concise presentation that summarizes a complex problem or issue

and presents the ideas, and one or more solutions and their underlying rationale. White papers are usually aimed at individuals or organizations that need to make a decision about how to solve the problem at hand. Although white papers need to be approved by the organization that issues them, they are usually not subject to peer review or edited in the same way that a scientific paper published in a journal is reviewed and edited before it is published.

In contrast, "technical reports" are documents that researchers write to detail the methods and results of their research. They are usually written for the agency or other organization that sponsored or funded the research. Technical reports are not normally subject to peer-review and usually have to be substantially rewritten before they can be published as a white paper or submitted to a peer-reviewed journal.

Finally, "preprints" are in between technical reports and peer-reviewed publications. Like technical reports, preprints are not peer-reviewed. But preprints have been reworked to the point that they're almost ready to be submitted to a journal for additional evaluation and external review. Researchers post their preprints online to discipline-specific preprint servers (also called archives or repositories) where other researchers can access them, read them, and comment on them (ArXiv is the oldest preprint server, but most STEMM disciplines now have their own preprint servers). Researchers use these comments to improve their papers before submitting them to a journal. Preprint servers also assign a digital object identifier [DOI] to the preprint. Once it has a DOI, a preprint can be cited like any other article. When credit for a discovery is contested, the time-and-date stamp associated with the DOI also can be used to establish priority for research findings.

3. There are important but often unappreciated or poorly understood differences between a salary and a stipend. A salary is paid regularly (usually weekly or biweekly) to employees based on work done (e.g., as an hourly wage) and is taxed. A stipend is usually paid in larger blocks (monthly or for the entire project) to interns or trainees and may or may not be taxed. More importantly, payment of a stipend implies a mentor-mentee relationship, whereas salaries paid in internships imply an employer-employee (i.e., managerial) relationship.
4. One of the most comprehensive web-based aggregators of post-graduate international research and study opportunities is https://scholarshiproar.com/.
5. The Association of American Medical Colleges maintains a database of available post-bacc progams and opportunities.
6. For example, the British Council maintains listings for non-degree post-graduate opportunities in the UK. A broader online resource is at postgrad.com.
7. Many graduate students in STEMM work as "teaching assistants," grading exams and running lab sections. But teaching assistants rarely participate in developing the syllabus or course material for a class. However, many universities will be receptive and supportive of proposals for new or one-time classes developed and taught by graduate students. If you do have the opportunity to develop and teach your own class as a graduate student (or post-doc), that can give you an edge in the competitive market for academic jobs.

8. The goal of "basic" research is to add to the world's body of knowledge. In contrast, the goal of "applied" research is to solve a particular problem. In general, far more money, time, and labor is targeted toward applied research than to basic research, but the distinction between the two is actually quite fuzzy.
9. The paper by Carson et al. is a thorough guide to applying to graduate programs. Although written by and for ecologists and evolutionary biologists, the same suggestions, guidelines, and rules apply to graduate programs in any STEMM field.

Notes for Chap. 9

1. For a further discussion of the different attributes of supervisors, managers, advisors, and mentors, see Sect. 1.1 and Fig. 1.1 in our companion book, *Success in Mentoring Your Student Researchers.*
2. This is our own definition of mentorship and it frames our companion book, *Success in Mentoring Your Student Researchers.* We go through the many detailed elements of mentorship in that book.
3. In her discussion of different types of "impostors," Valerie Young writes that "the fact that everyone else sees [you as] a highly capable individual where you see [yourself] as an inadequate fraud, is a pretty good indicator that you operate from a competence playbook that bears little resemblance to reality."

The manufacturer's authorised representative in the EU is Springer Nature Customer Service Centre GmbH, Europaplatz 3, 69115 Heidelberg, Germany. If you have any concerns regarding our products, please contact ProductSafety@springernature.com

Printed and bound by CPI Group (UK) Ltd, Croydon, CR0 4YY
01/10/2026
02225839-0005